CALL THE FIRE BRIGADE!

The author as a Station Officer in the late 1970s, alongside the wheeled escape ladder, wearing the Proto oxygen breathing apparatus

Call the Fire Brigade!

Friendships, Fire and Fear

ALLAN GRICE

EDINBURGH AND LONDON

First published in Great Britain in 2012 by
MAINSTREAM PUBLISHING COMPANY
(EDINBURGH) LTD
7 Albany Street
Edinburgh EH1 3UG

ISBN 9781780575667

A catalogue record for this book is available
from the British Library

Printed and bound by
CPI Group (UK) Ltd, Croydon, CR0 4YY

1 3 5 7 9 10 8 6 4 2

Contents

Author's Note

Having spent 20 years as a fireman and officer in the highest risk areas of the London Fire Brigade during some of its busiest periods, I have experienced much of life and death and the general harshness of our existence. Moving from West Riding County Fire Service in 1972, anxious to feel the excitement, as I felt sure it was then, of a metropolitan fire service, I joined the London Fire Brigade and so began my initiation into the brutal reality of life in a teeming city, brimming with danger and poverty.

Often the hopelessness and desperation of life in areas where we responded to emergencies was palpable – even as I crawled, face to the floor, trying to escape the dense smoke of a deadly fire. This was especially the case when responding to 999 calls in the cosmopolitan East End, that district of mean and moody streets, meths drinkers, teeming tenements, densely crammed workplaces and the giant warehouses and wharves of the London docks, then in the final years of their once distinguished and colourful life.

Over the many years that I worked on the front line as a fireman, I kept a brief note of the fire and non-fire emergencies I attended. I wanted to be able, if I lived that long, to one day look back upon those days and nights spent protecting the people of the capital, and their properties and livelihoods. By drawing on selected incidents and situations, I have described the real-life drama and humanity of the world of fire and rescue of over 40 years ago: of the sadness felt by crew members at the loss of life, and the elation experienced when lives were saved.

Why did I write these accounts? There were a number of motivations. Many people, on learning of what I did and where I had served, asked me questions about life within the fire service. This got me thinking that there would be others out there also curious to know. On top of this, as far as I could see, there were very few accounts in existence that recorded the perspective of someone who had actually been there, in the thick of it, back then.

The 1970s was a decade in which there was a massive sea change in attitudes to life and society. Some of these were a natural result of progress and were welcomed, but within the fire service some of those new ways of thinking would herald the end of an era. With the introduction in 1974 of legislation regarding health and safety in the workplace, the old 'firedog' mentality – and the cult of the 'smoke eater' – was dealt a blow.

Over time, as the legislation took hold, some with command and control responsibilities became frightened of transgressing the safety laws, to the extent that their natural and developed professional instincts for selfless action towards those needing help were throttled. The UK fire service has always prided itself on its overall high standards of personnel safety in what is potentially an extremely hazardous occupation. Given the thousands of dangerous situations dealt with, the safety record statistics of the UK fire service in terms of death and injury are excellent. Strong discipline, firm but fair management and experience-based leadership always played a key role in the safety of fire and rescue crews of my era on the front line.

Those of us who, because of the lack of breathing apparatus, had to endure vile smoke and fire fumes will always be grateful that those who came behind us did not have to suffer such a hazardous exposure. Health and safety requirements today ensure high levels of personal protective equipment and training so that firefighters are better able to cope with all eventualities, and this truly is a blessing. The concept of personnel safety is a noble one, indeed; however, if the regulations become unbalanced in their application – and I have seen this happen – a distorted preoccupation with the safety of emergency service personnel starts to inhibit overmuch the essential human instincts to place one's own life on the line to save those in peril. Mollycoddling – wrapping people up in thick cotton wool – is a distorting process that plays its part in making individuals averse to risk. It is at this point that serious questions must be asked of legislation.

For me, the sliding pole, the ear-splitting sirens and blue beacons were there to help us get to the scene as rapidly as possible. Both Pumps were usually crossing the threshold in not much over a minute during the wee small hours and at the incident within four more. No other service could beat that. This was a world away from what arose in later years, when this vital rapid response by our sister services was halted within high crime areas until police escorts could arrive to help ensure the rescuers' safety whilst those awaiting help may have lain dying, thanks to an unintelligent application of health and safety legislation principles.

London Fire Brigade is world renowned for its high courage and fire and rescue effectiveness. There is no doubt in my mind that this reputation was grounded in the massive practical experience gained on the busiest inner city fire station areas in times of peace and war when the character, attitudes and traditional decency of people in all walks of life were markedly different from today.

So my accounts are about another time and about a type of character that rarely exists today. My old 'guvnors', men who had been tempered and forged in the searing heat of the Luftwaffe's Blitz raids on key locations within this country, as well as serving on the front line in Europe, were from an era where military-style discipline and hierarchy could mean the difference between life and death.

Although it is true that the overall numbers of emergency calls received today are generally higher than was the case 30 or 40 years back, the numbers of serious life-taking and life-threatening types of fires, and the numbers of major property fires responded to in this century, are far fewer than those attended by London fire crews of the years described in the accounts which follow.

Fires were dealt with without the level of personal protective equipment provided today. Perhaps the biggest shift seen, though, over the years since my accounts, has been that related to the operational strategy employed. Then the approach was one in which officers in command insisted that we 'got into fires' at all costs in order to save both life and property; today, safety laws temper and, some would say, tightly shackle such an aggressive approach.

This is a no-punches-pulled, warts-and-all account that, in some places, is not for the squeamish. It gives the truth of what it was really like to be a fireman in the 1970s – that turbulent decade of massive political unrest.

Above all my accounts allow you to walk within my world as a firefighter – to feel the tingling on your ears that alerts you to the heat, urging you to get out of the building; to see the charcoaled remains of women and children with no escape from fire; to experience the elation of saving a life; to smell that unmistakeable odour through the Pump window en route to a call – the cocktail of burning wood, paint, fabric and rubber all in one – telling you to take a deep breath and face up to what's coming.

* * *

By its very nature, the fire station is a place where, especially on the 15-hour-long night duties, crews talk over the jobs that they have been involved in. It was through these conversations that I was able to see incidents from the perspective of the other firemen on duty (female firefighters were not recruited into the London Fire Service until 1982, so in my day it was a purely male contingent) and am able here to relay their experiences. In addition, having had the great privilege of not only being a front-line fireman but also, eventually, a watch (shift) manager, I am able to convey what I personally experienced and also what I know those who were my guvnors and mentors during my first years in London would have been feeling when responding to and effectively dealing with the hundreds of emergencies we attended back then.

While my accounts are based on notes taken at the time, many of the most dramatic responses – and some, like the dreadful

Moorgate tube disaster of 1975 – were so horrific that the memory of them is seared onto my mind like today's digital technology can burn an image onto a CD. I can recollect scenes in intimate detail still of some of the more horrific rescues we carried out.

As not all of those whom I worked with, and their children and the public we served, have passed on, in order to respect their privacy I have used pseudonyms for all of the actual personnel I was so privileged to work alongside; furthermore, some situations and locations will still be emotionally sensitive, so I have omitted some street names. None of this alters the fact that these are true accounts of my life within the Big Smoke in those years before the nanny state with its throttling, European Commission edicts on health and safety at work altered the way in which we had operated so effectively for so long.

Chapter 1

Death in the Small Hours

'*Fire in tenement – multiple calls being received.*' Our officer in charge, 'Biff' Sands, is reading the slip through sleep-bleary eyes as I and the rest of the watch plunge down the sliding pole and look up through our own fuzzy vision to the coloured indicator lights fitted above the station doors that indicate who will be responding. Green is for the Pump with its crew of four; red is for the Pump Escape, with a normal crew of five. Both lights are glowing and the diesel engines growl like chained guard dogs sensing an intruder and raring to go. They fill the air with pungent exhaust smoke, through which the light from the fast-spinning electric-blue beacons eerily reflects.

Blue Watch is the duty shift. Biff is leading it, in charge of the Pump. The guvnor of this watch, he is distinguished by a white helmet with a narrow encircling black band. He is supported by his number two, Sub Officer Jack Hobbes, and his number three, Leading Fireman Dick Friedland. These three manage us six firemen, whose ages range from a callow nineteen to the grizzled early fifties.

We quickly pull on helmet, boots and leggings, then a silk neckerchief, which protects against hot embers getting past the collar of the cloth tunic. Finally, the belt and axe pouch is secured.

The Pump, with its Dewhurst ladder, which can be extended to 30 feet, roars out hard on the heels of the Pump Escape, the brigade's principal rescue ladder, a one-ton wooden beast with four-foot-diameter wheels, enabling it to be swiftly positioned and then hand-wound to a maximum height of 50 feet. Less than 60 seconds have gone by from the first urgent clamour of the call bells to our leaving the station, now silent and deserted like the *Mary Celeste*, with only kicked-off shoes lying on the red-tiled floor as proof that we were there.

Such is the importance that the London Fire Brigade attaches to the speed of response that a senior officer can turn up unannounced by night or day to implement a test turnout by actuating the call bells and timing us with a stopwatch. Woe betide crews that are sluggish. No matter how deep a man might be sleeping in the iron army-type bed provided, once the call bells clamour he is expected to be awake and on the appliance in seconds. You never know if the call is for real or if a cunning and impassive-faced officer is behind the rude awakening. Turnout competitions are held and the rivalry between stations to secure the trophy for the swiftest is intense.

No matter how much some men might curse them, the motive behind these test turnouts is noble – long experience within the high-risk inner city has hammered home how the saving of seconds can be vital where persons are trapped by fire or hanging by their fingertips on a ledge where they have crawled to escape the searing flames and choking, superheated smoke.

The mean and moody East End street where we have been called is in what at the time would have been called a dodgy district, the kind where coppers patrol in pairs. The street is thick with buildings: five-storey tenements crammed in next to furniture and garment makers; printers, plumbers and pawnbrokers side by side with wine and spirit merchants, timber and builders' yards, and a host of other commercial premises in the heart of this seedy enclave.

The cab window next to Biff, our guvnor, is open a few inches in spite of the bitter cold of this January night and the job can be detected even though it's several streets away. Building fires create an unmistakeable odour. It is a cocktail of burning wood, paint, varnish, fabric and rubber all in one. No matter how many hundreds of times it enters the nostrils, heart rates quicken and the adrenalin rush is full bore with the nervous anticipation of what is to come.

Both pumps career round the bend into the street and immediately run into the heavy hot fog of thick brown smoke that has all but obliterated the sodium street lamps.

There are long tongues of yellow and red flames shooting fiercely skywards from several windows at the fourth and top-floor level of a grey tenement. A burning timber window frame is on the pavement, having fallen from on high.

'We've got a right bleeding goer here, boys,' shouts Station Officer Sands.

As we squeal to a halt, a shocking sight confronts us. A man is impaled on a set of spiked iron railings that separate the street from a basement area below. It is clear he has jumped, chancing his luck from fifty feet rather than being burned alive, but with over six inches of pointed steel sticking out of his chest, it looks as though his luck wasn't in.

The noise of the fire and the roar of the arriving pumps have awoken residents of the tenement and other dwellings in the street. Some 30 or 40 folk are on the pavement opposite the burning building. Some are wrapped in blankets, with striped pyjama legs just visible above carpet slippers; others are fully clothed, their necks craned back as they look up at the inferno, hands on mouths with shock. A couple of police officers are at the scene: a six-foot-plus PC with the build of an all-in wrestler and a much shorter WPC at his side. Both are trying to help the unfortunate speared from back to chest by the railing spikes.

I can see Biff calling on his vast experience to inform his decisions. He needs to work out how many men will be needed to best save life and property. With three pumps automatically dispatched to all building fires within this powder keg of a district, and with a fourth on its way because more than one 999 call has been received, he knows that the best part of twenty men will soon be battling this blaze. Adequate for the time being.

Jack Hobbes is detailed by Biff to check if any fire is visible at the rear of the tenement and if anybody is visibly trapped at windows.

Biff establishes from a man who says he is the caretaker for the block that a family of three occupy the top-floor flat from which the flames are billowing. He thinks that the person on the railings is one of them, but the other two occupants are nowhere to be seen.

Once Biff hears this, he instructs the Pump's driver to send the priority radio message, '*Persons Reported*'. This will result in Fire Control ordering an ambulance, but the pool of blood on the pavement means it's probably too late for this man and all efforts will be directed at firefighting and rescue.

Fellow Yorkie Barry Priestley and I have been instructed to rig in BA. Once the third pump arrives, a further pair of BA men will be similarly instructed and they will back us up. Since time is crucial to fire suppression and victim location, Dick Friedland and crew member Ricky Tewin have already entered without BA. They have taken in a jet and one will operate this whilst the other begins a preliminary search for the missing persons before the BA crews relieve them.

Unless a fire is so well advanced that we are forced to tackle it from the street, the drill is to drive the fire and heat back outwards through the windows rather than pushing it inwards, which can spread the fire and lessen the survival prospects of anyone trapped.

After Jack confirms that no persons are trapped at the rear, Biff makes his way up the stone staircase to the fire floor. Officers like him don't ask their men to go to places they won't go themselves, so he steps around the hard red snake of hose and meets the heat and smoke at the top-floor landing. It is so hot, so thick and so pungent that it forces him onto his knees, making him curse at its awful ambience. It is fiendishly hot, but his earlobes have not quite begun to tingle with the heat. He knows that when they do it's an urgent warning for Dick and Ricky to get out fast before the whole apartment erupts into a deathly lung-searing fireball. But for now he pushes on, confident that he'll know when things are getting too hot for safety.

His knowledge is grounded in advice handed down by his mentors of years back, when he was still a young fireman; mentors who were greatly experienced men, old 'firedogs', as they were known – long-serving, seasoned and very competent fire officers, many of whom had earned their spurs and that title fighting the conflagrations of the Blitz. He has augmented that advice with 20-odd years on the ground in the heart of what are some of the capital's highest fire-risk districts.

There is a thin vein of air just above the floor. Biff is now flat on the fifth floor landing, breathing this in, his white helmet almost invisible in the gravy-thick acrid smoke. He can hear the crashing noise of the jet being operated by his men, who are taking a lot of punishment fighting the flames in that cruel interval before others wearing BA arrive to relieve them.

In the days before there was a BA set for every crew member – something that didn't happen for quite a few years after my transfer to the capital – those who were to wear BA for the shift would be taken from a roster and announced at roll-call. When you responded to a job where it was clear that breathing apparatus would be required, you only put the set on when your guvnor ordered, 'BA men get rigged!' or 'Don your sets and start up.'

To rig properly in BA – the closed-circuit 'Proto' apparatus – was quite a lengthy process. The set has a harness and breathing bag that go over the shoulders and the oxygen cylinder is in the small of the back, along with a brass on and off valve in the form of a wheel. Another two controls allow you to either send more oxygen into the chest bag if the work is heavy or to permit an overfilled bag to vent the excess. The latter is frowned upon, as you are getting rid of oxygen that might save your life should you become trapped and capable of operating the valves. A set of rubber-framed goggles keeps the eye-stinging smoke out and a mouthpiece secured to a personal head harness is placed between the teeth. A set of nose clips keep you from inhaling the deadly fumes.

It is because properly rigging takes time that Dick and Ricky went straight in under that vile smoke to begin their combined search and firefighting efforts: seconds saved can be the difference between a victim being dead or alive.

I quickly tighten the body belt, take out my personal hand-sewn head harness, which I keep inside my helmet, and position it ready to attach the mouthpiece via two D rings. Mouthpiece in, I crack the valve and feel the chest mounted bag engorge. I secure the nose clips, pull down the goggles and put my helmet on, then hand in my name tally to the Pump's driver.

With Barry Priestley, I make my way to the fire floor, wondering as always whether we will get out in one piece, recalling a recent fire at a Chelsea restaurant where two firemen had died, the pure oxygen in their BA sets apparently having intensified the burning of their air passages. But these thoughts cannot be dwelled upon.

We reach the top floor, where Biff is on his knees, barely visible amid the smoke, the awful product of burning furniture and floor coverings. We enter the first floor and such is the heat we are forced onto our knees. Barry's thick ex-coalminer bunch-of-bananas fingers grasp the red snake of fire hose which Ricky and Dick have taken in. The crashing of its jet is loud as they beat down the searing flames and so dense is the smoke and so terrific the heat that the pair must be on the point of collapse. Though the mouthpiece renders clear speech impossible, Barry mumbles, 'Get out – get out now!' Barry takes the nozzle, swirling it around to lower this awful heat – it's far worse than being inside a gas oven on high – while I start a search of the apartment.

I know from personal experience that Ricky and Dick will have by now crawled out, coughing and retching violently on hands and knees, their faces red as lobsters and glistening with the intense heat and strenuous effort. Their noses and mouths will be caked in the snotty mucus caused by ingesting the foul smoke.

I start my search for the missing persons, following a systematic plan. First around the edges of the rooms, then diagonally across in sweeping arcs, with hands feeling for victims, probing under tables, on sofas, and on and under beds. I then check inside the bath tub and cupboards, under the windows and behind the doors – the last two places being where we often find victims who have made a desperate effort to reach safety and clean, life-giving air.

Only ten minutes have gone by since our arrival but already most of the fire has been suppressed. The terrible smoke is now thinning rapidly and the skin-searing heat is falling in temperature by the second.

Biff can now afford to stand up, even though the atmosphere in an upright posture is still a lot hotter than on the ground. At least he is able to make out Barry handling the jet and myself and others searching earnestly for the two occupants still unaccounted for.

There is a grunt from one of the other BA men instructed to assist us with the search for victims, his speech gagged by the mouthpiece and nose clips of his oxygen set. He signals for the Station Officer to come over. He points his lamp beam to the floor and onto a steaming pile of what looks like the debris of plaster, burned ceiling laths and the remnants of incinerated furniture.

It is neither of these. What it is sadly is the partly cremated remains of the two missing residents. They must have been asleep in a double bed in the front room, where those wicked flames were issuing on our arrival. The heat has been such that the skin on their torsos has split like that of a sausage that has not been pricked before going under the grill.

As always when I see death, especially where the terrible burning makes it absolutely impossible for the victim to be alive, my mind thinks of the absolute irreversibility of that state. For me, it is like a 'frozen in time' image – 'the decisive moment', captured by French street photographer Henri Cartier-Bresson. It is indeed a sobering sensation, reminding you to savour the fresh air and your good health while you have them.

A radio message is sent and the matter-of-fact wording reduces the terrible tragedy to the details of an official report:

> *Tenement of five floors, 60 feet by 40 feet. Twenty per cent of top floor damaged by fire, heat and smoke. One male jumped from fourth floor to street before arrival, apparently dead. One male and one female found in front room on fifth floor, apparently dead, all awaiting removal. One jet, one hose reel, four BA.*

The next hour is spent ensuring the fire is fully extinguished. The apartment is a blackened, steaming mess. Debris of furniture, timber, plaster and lath is turned over and damped down so that there is no prospect of it reigniting. But it is done without disturbing the charred human remains that might contain vital clues as to the cause of the blaze. Fatalities mean that the incident becomes a matter for the police and coroner and that at some future date those in the Brigade who discovered the bodies will be called to court to report the facts of their involvement.

The gas and electrical authorities are called to verify all is safe with their equipment and a senior officer from the brigade's Divisional HQ has come on, as it is the policy for one to attend jobs where lives have been lost.

The detailed investigation into the cause of the fire reveals that a paraffin heater had been knocked over by accident. This had flared up and the leaking fuel had turned the apartment into a furnace within minutes.

Two hours after our arrival at the fire, both pumps are back in the station. Biff Sands is writing up his reports and the rest of us are busy re-stocking hose, cleaning and servicing the BA sets and polishing the red bodywork until it gleams. All of this to ensure that we are ready for the next shout for help from wherever and whenever it might come in this mighty metropolis of seven million people, with its thousands of premises, large and small – and the calls never stop coming.

Chapter 2

Beginnings

When I was 13, I happened upon a book by former inner London fireman John Anderson. *The Fireman*, published in the 1950s, described Anderson's career before, during and after the war. The book's opening words stirred something within me, something that somehow integrated with earlier medical ambitions of mine, and my wish to do something worthwhile with my life:

> Bells, so often it begins with bells, and then round the corner two gleaming red fire appliances come into sight and you have a brief glimpse of helmeted men inside as they charge past . . .

I was too young on leaving school to apply to the fire service, so instead I secured an apprenticeship in one of the city of Leeds' most famous colour printers, Alf Cooke's in Hunslet. The mid-1960s is remembered by me, from the standpoint of today, as an important staging post, a springboard from which I would make my 'leap for life' (those three words were to be relevant to my journey, too). Perhaps when you are still in your teens it is always a good time to be around, on that exciting threshold of your future; the truth is that the mid and late 1960s were exciting and memorable periods for so many.

The 'Fab Four' had only recently started out on their journey.

So different was their music from what had gone before. The haunting lyrics, and their captivating and earthy personalities, soon to charm the whole world, were also ringing the changes of a society that only two decades earlier had been suffering the privations of war. Leeds was, of course, the birthplace of Jimmy Savile, the former road-racing cyclist and DJ. He could often be seen charging through the Headrow, a main street in the city centre, on his bike, his platinum blond locks flowing in the air and his 'Savile's Travels' motorhome in front. The same air also seemed to be carrying a wind of change in politics, as PM Harold Wilson, a son of the textile town of Huddersfield, made his speech about 'the white heat of technology'.

It was definitely an exciting time to be young and to experience what this new age would bring. What it brought for me, although I did not, of course, know at the time where it would end, was a series of events that linked in to those opening lines in John Anderson's book.

It was whilst I was on the shop floor of Alf Cooke's massive printing works one winter's morning in the mid-1960s that something happened that helped consolidate my ambitions for a life of fire and rescue work. The staccato peal of fire engine bells announced the urgent arrival of two pumps from the local station of Leeds City Fire Brigade. They had been called to deal with a mysterious haze of smoke that had appeared in the main press room. As three or four of a crew entered, resplendent in their gleaming black helmets, boots and double-breasted Navy-style fire tunics, Eddie, my press assistant, who had seen active service in Italy during the war, said, 'What a fine body of men.' Back then, the sight of uniforms was reminiscent of a not-very-distant past when so many, like Eddie, had donned uniform either as regulars in the 1939–45 war, or in the Korean conflict or had completed national service, which had only ended a few years earlier. As such, the general public were much more willing to respect the courage and sense of public service displayed by firemen and police officers than would be the case as the years went by, when a sinister and unhealthy resentment of anyone in authority took hold amongst many people. For me, the presence of those smartly turned out firemen brought back the words and images in Anderson's little book.

However another much more dramatic event hammered the nails into the coffin of my printing career, provided that I could unlock the tight shackles of my five-year-long apprenticeship – a not inconsiderable feat in those days. I was on the second of two night school classes, part of a City and Guild's course in lithographic printing. It was about eight in the evening in early summer and I was at a desk overlooking the main street down into central Leeds. I was being sent to sleep by a lecture on the constituents of printing ink when an urgent ringing clamour aroused my interest. Anderson's line –'Bells, so often it begins with bells . . .' – was again triggered in my mind. I looked out and saw the Pump Escape and Pump from Leeds Central fire station race past. Within the next few minutes, more bells and the wail of a New York-style wailing siren, as two more machines roared by.

With the lecture over, I started the long walk to the central bus station. It was just becoming dusk but about a mile away there was a huge, sinister-looking black smoke plume reaching hundreds of feet into the darkening sky. I reasoned that was where those fire appliances must have been heading.

'The Calls' is an old Leeds street. Today, it is a district of trendy apartments, nightclubs, hotels and bars, but back then it was a heavy commercial district of warehouses abutting the black waters of the River Aire. It was one of these warehouses, which I later learned was stuffed with highly combustible raw wool bales, that was fiercely and spectacularly ablaze. A baby-faced bobby, who was probably not long out of recruit training, was at the top of the street, doing his best to prevent the mounting crowds attracted by the smoke and fire bells from getting too close. But I managed to circumvent his restriction via a back alley, which got me almost into the affected street. I could not believe the sight of the angry rolling red and yellow flames that were billowing from the highest windows and roof of this multi-storey redbrick Victorian warehouse. Such was the size of the fire that the rapidly falling night appeared in the surrounding streets like a bright summer's day.

Atop a Turntable Ladder, which must have been a good 70 feet high, the fireman looked like a toy soldier silhouetted by the lurid flames. A silver rod of water from his hose was being played on a slated roof on a premises next door in an effort to prevent the massive waves of heat being radiated, igniting first this building, then the next. The cloud of steam that arose when he played the jet elsewhere indicated how hot those slates were.

As I stood in my doorway, almost transfixed, the night air was rent by the bells and sirens of more and more reinforcing appliances, the name placards on their lockers indicating they had been dispatched from such neighbouring towns as Wakefield, Rothwell, Bradford and Dewsbury. Stentorian commands in the hard consonant dialect of West Yorkshire were being rattled out like bullets from a machine gun from a shortish man whose helmet markings indicated that he was brass.

A glance at my watch showed if I didn't pull myself away from this high drama, I would miss the last bus and that would entail a 14-mile walk home – not good when you have to be up at five the next morning to ride the double-decker back into town to start a shift at 7.15. On the bus, which I just caught before it pulled away, I went over all that I had seen and by the time I was home I knew the die had been cast. I wanted a life coloured by fire, smoke and rescue work, not the rainbow spectrum of printing ink.

Although there were not the same number of applicants back in the 1960s for the fire service as there are today, I was still up against stiff competition when I learned of two vacancies in the Wakefield City Fire Brigade. Back then, because there were so many employment prospects, and because a fireman's wage was so paltry, you would normally apply for such a potentially hazardous job only if you were dead keen on the work, or because it was a last resort after failing in other job applications. Whatever the reason, the senior brass interviewing still wanted to recruit the most suitable applicant, so they would be looking for evidence of knowledge of the fire service, plus a good attitude. The fascination I had for the work, first stimulated by John Anderson's book, meant that I was well up on what the service was about. I must have been convincing, or the other applicants poor, because to my great delight I was offered one of the two posts.

Even though the brigade was bang in the centre of the county's headquarters town, it had no connections with the West Riding in terms of fire protection. You see, up until 1974, when local government boundary changes took place nationally, the UK fire service consisted of a mix of county council and county borough brigades. So although surrounded by West Riding county fire stations, such county boroughs as Wakefield, Leeds, Bradford, Halifax and Dewsbury were their own entities with their own budgets and hierarchies. To be a high-ranking officer in a county borough was to be a big fish in a little pond, whereas the same rank in the much larger county brigade was a case of a smaller fish in a much bigger pond.

But before I could accept the job offer I had to get out of the iron-clad indentures of my apprenticeship. Much to my delight, with some persuasive talking, they released me and I was allowed to leave. With luck, and by successfully getting through my recruit training, I would be in an occupation in which I was not hemmed in by walls and a clocking-in machine.

I did complete the training – it was the nearest I would ever get to being in the forces. Most of the recruit instructors were former military with active service and, being serving City of Leeds firemen with a fair bit of city centre experience under their belts, they instilled in me a sense of fair but no-nonsense discipline that stood me in good stead in my later career.

My two years in this one station outfit at Wakefield taught me that when a fire station does not deal with very many emergencies, the potential for cynical and lazy attitudes develops. (In later years, a grizzled London senior ranker would state that the happiest stations were always the busiest.) So when I learned of a vacancy in the county brigade, I applied to transfer, reasoning that the greater opportunities afforded by a large number of stations would give me the practical experience I sought. But always at the back of my mind it was my ambition to one day become a London fireman.

The reason I sought this 'shop floor upwards' experience was connected to the fact that by dint of rigorous study I had managed to get two of the statutory promotion qualifications under my belt before I was 23. (To achieve permanent promotion, a fireman had to pass a written/practical national examination to qualify for the first three service ranks. In 2004, these qualifications were made obsolete and replaced by a psychometric test-based method.) My inner drive and tenacity were perhaps inherited from two coalminer relatives on my mother's side who had hewn the black gold from the deep mines of the Yorkshire coalfield. Such results when you are so young and look so boyish, however, can create resentment and jealousy, especially in the minds of those who are past the first flush of youth and find exams difficult. I was determined, therefore, not to attract the pejorative label of 'bookman' and my transfer to the West Riding County Fire Service, and a posting to Castleford, home to a top-flight rugby league team known to the locals as 'Classy Cas', was hopefully another step on the road to my gaining more experience.

Although the station was not particularly busy, I attended a lot more incidents than at Wakefield. When I look back, it was one that ironically occurred while I was off duty that, I believe, became instrumental in the way I carried out my job in later years. It not only consolidated my wish to seek a huge practical experience but also played a part in the way that I would eventually manage the safety of men I was in charge of. It hammered home to me life's fragility but also the importance within an emergency or military service of shared belonging.

I had been on nights and was driving back to the two-up two-down mid-terrace my wife Carmel and I had recently bought in a little village off the A1 Great North Road. Back then you came on duty and went off duty wearing your undress uniform so you were easily recognisable as being in one of the emergency services. On a busy road, infamous for road traffic fatalities, and near the entrance to one of the many collieries of the district, I saw a car half in a ditch and impacted into a telegraph pole. Although I would have stopped anyway, my being in uniform seemed to compel me to see what had happened and, if appropriate, render what first aid help might be needed.

Sadly, it was too late for first aid. The driver, who it turned out was an Asian GP, had collided with the telegraph pole and, it later transpired, been killed instantly from a broken neck. I had not seen death before and because I was not with a team of crewmates, I felt the shock of this sudden ending of life – the sudden awareness of the sword of Damocles that hovers above us all – more than I would have otherwise, as part of a fire and rescue team. The mutual support, the so-called *esprit de corps* that lies at the heart of an effective emergency service unit, was not there because I was outside its sustaining force. I remember so clearly how the GP's skin was an ashen grey and how the still open eyes stared lifelessly at me. In the succeeding years, I would witness death many times, often amid scenes of devastating carnage, but as horrific as some of those fatalities were, they were easier to deal with than on that first occasion. Of course, one becomes case-hardened – a mental callus forms over time which protects – but I think one of the most important contributors to the efficacy of an emergency squad, be it in a hospital, on a battlefield or as a member of a mountain rescue or lifeboat, is this support that crewmates give one another.

That young doctor's death was another consolidating factor in my eventually always appreciating the guvnors who 'ran a tight ship' but knew when to let their men relax. The former applying, of course, to every emergency response; the latter during stand-down time in between calls on the long night shifts.

I constantly wrestled with the idea of applying to transfer to London, but my having passed two important promotion examinations must have been noticed by someone at BHQ because I was promoted at the young age of 24 to Leading Fireman. I was posted to the inland port of Goole, which promised much action in its wharves and quaysides, but it was very quiet and my mind was looking south yet again.

I had reason to attend Brigade HQ one cold January morning and in the mess room I picked up a copy of the *Sun* newspaper. In it was a report on Prime Minister Ted Heath's signing the first part of a Common Market document. My eyes were then drawn to a dramatic full-page image, captioned by those three words: 'Leap for Life'. It was a photograph of a fatal fire that had occurred on a Sunday morning in a flat above a post office on London's King's

Cross Road. Crews were manoeuvring the wheeled escape ladder, two ambulance men were tending to two young people who had leapt out of the top-floor window and two other firemen were preparing to work a hose into the building. That chance pick-up of that paper crystallised everything. I knew for sure that, as big a move as it would be, I would apply for a transfer into the London Fire Brigade.

As I write these words, I can look over and see on a wall a framed copy of that photograph, which played such a part in my own 'leap for life' as I began the process that led to my adventures within the Big Smoke.

Chapter 3

Whiskey in the Jar

I'm quite cold, standing on the platform of a surface station of the London underground. My usual mode of transport to the fire station, over 15 miles across the congested roads of the capital, is by car. Unfortunately, after ten years and 100,000 miles, the gearbox has decided to call it a day. As a consequence, here I am, at 4.30 in the afternoon, taking the tube. Fortunately, it's not too far to walk to the station the other end, where the first of my two night duties begins at six.

After waiting for a good 20 minutes in the chilling breeze, I am glad to see the red snake of the train pull in. It's good to feel the relative warmth of the carriage as I grab one of the few empty seats on what is a train packed with mostly homeward-bound commuters. No matter how many times I ride on the underground, the unique association with London that these trains and stations evoke in me is always there – their smell, their sounds and that tingling anticipation of new experiences that the capital provides.

Perhaps it is the abundance of ever-changing advertisements. Those long, curving strips underneath their glass and acrylic cases, with their stainless steel frames, so many of which evoke the London scene of commerce, leisure and endless opportunity.

Above all, perhaps the unique atmosphere that I feel in London comes from the tightly crammed, cosmopolitan crush of people within the confines of the clattering carriage.

To travel on the London underground during the peak hours is to travel amongst so much of what helps to make the capital what it is: a huge, sprawling, bustling, rushing, teeming fleshpot; a cosmopolitan container of people that daily disgorges its human cargo across the length and breadth of the city.

On alighting, this thronging mass makes its way to countless destinations for countless reasons and purposes, each with a specific motivation . . . most honourable, some not. But wherever this teeming throng of people go, there is no place within this conurbation that falls outside the embrace of the brigade and its fire stations, the crews of which stand their safety vigil by day and during the long watches of the night.

Wherever masses of people live and work, the risk from fire and other calamities will exist. For myself, there is therefore nothing like a crammed full tube train to best remind me that it is within the largest conurbations that the greatest risks to life exist.

I step off the train onto the platform deep below the mainly mean streets of the inner-city suburbs where my fire station is located, wondering what act of human anger, weakness or mistake might occasion an urgent call of the turnout bells on my 15-hour-long night duty.

'Busy day, Mart?' I enquire of Marty Molloy as I enter the station via the open doors of the appliance room, where he stands looking on to the street. Marty is a 20-year veteran of the brigade and one of the most respected firemen around.

'Nah, couple of mickeys and a "wash petrol off roadway" after a two-car shunt outside the station. Still, if nothing else, I won't be falling asleep with fatigue tonight,' he answers, this last remark a reference to his part-time job on the door at a local dance hall. Marty had been a useful light-heavyweight boxer as a young bloke. Today, he tips the scales at fifteen stone; at just under six feet, with powerful shoulders, close-cropped dark hair, flattened nose and square jaw, he is a good visual deterrent to most would-be troublemakers fuelled by drink.

'Well, perhaps it's our night for some action, Mart. Just hope you don't get too much, if you know what I mean.' I then go up to the dormitory locker room to change into my uniform, ready for the roll call, due in 20 minutes.

'Looks good, Paddy. What's in the pan tonight, buddy?' asks Ricky Tewin of our 'chef', as he strolls into the first floor galley kitchen, following our couple of hours out in the drill yard.

Each watch on London fire stations had a mess manager – it was usually a job that certain fellows who fancied themselves in the cooking department sought. During weekdays, most stations employed a civilian cook, usually a woman, but it was still the mess manager's responsibility to buy in the food for meals using the monies that each man contributed. However, at weekends and on night shifts there was no civilian cook and at these times it was the chef's job to rustle up a satisfying meal. This meant he would be occupied straight after the six o'clock roll call, so he was excused drills on nights – the more cynical would say this was an incentive to take on the role!

'Sausage and mash with onion gravy, followed by tinned oranges and cream, mate. How's that grab you?' Paddy Mulligan, a 14-year veteran of the brigade, replies.

'If it tastes as good as it sounds, it should be handsome, Paddy. You've missed your way in life. You could have been another of those famous television chefs.'

'What, and miss working with this bunch of reprobates?' he grins back, as he places some food inside the large refrigerator.

Later that evening the sausage and mash does taste good. I fork up another delicious mouthful, but before I can put it into my mouth, the lights come on, followed by the sudden clamour of the call bells.

'*Fire in derelict house. Pump only*,' the dutyman calls out, as we spring off the rubber pole mat onto the ridged red tiles of the appliance room floor.

Our neighbouring station is out on a six-pump fire in the West End, so we have been called to cover part of their ground. The derelict house is not far from one of the capital's main line railway stations and is in one of those dingy backstreets where you wouldn't feel safe alone at night.

When we arrive, a thin roll of grey smoke is coming out of the first-floor window of a derelict Victorian house. We jump down and pull off the hose reel tubing automatically.

It is an old horsehair-filled settee that is burning, either ignited by kids or by down-and-out vagrants. Within ten minutes, we have soaked it thoroughly, to the extent that no one will be able to reignite it for a long time. Within another 15 minutes, we have rewound the tubing and returned to the station. There, Barry Priestley and I top up the onboard water tank on the Pump.

The hose reels and water tank are the most used pieces of equipment. Probably some 90 per cent of all fires are extinguished using the reinforced red rubber tubing, which is tightly wound on two drums, one on either side of the pumping appliances. The onboard tank contains 300 gallons, but because the small diameter but high pressure nozzle uses a relatively low amount of water, this quantity can be used to excellent effect at all but those fires where a large body of flame is present.

The hose reel is the first line of attack on many fires. Even on the larger jobs, it can be used as a 'holding jet' whilst the large delivery hose is laid out and charged.

After a quick wash of hands and face, we are soon back up in the mess. The sausages have baked a bit hard in the oven's warming tray, where Paddy has placed them, and the gravy has congealed, but the oranges and cream are good, as is the big mug of tea poured from the huge metal pot.

I head up to the dormitory at about 10.30 and lie on the hard bed, intending to read a book about London's down-and-outs.

I must have been tired because when the lights and call bells arouse me just before a quarter to one I am still in my day gear and the opened book is on the floor at the side of the bed.

With London's homeless still in my head from my earlier reading, I rub the sleep from my bleary eyes, grasp the polished pole and drop swiftly down. The call bells still clamour and the blue haze of diesel smoke follows the roar of the engines on both pumps, which are revving loudly, ready for the charge through the early morning streets on which many of the night's revellers are still present.

We can see the smoke billowing out of a top-floor window from a good thousand yards away. As we charge down the high street, the electric blue sweeps of the rotating beacons bounce off the windows and walls of the many commercial premises of this district. Although well after midnight, the streets are still busy with cars and pedestrians on this Friday night. It is at this time that the pubs, clubs and dance halls do their best business.

I would guess that it had been one of the many passing pedestrians who first noticed the brown smoke issuing from the sixth-floor window. He will probably have looked around for a telephone kiosk to call us. I have seen fires off duty and know how welcome the sight of a call box is – you only hope and pray it is not vandalised, with the metal-reinforced cable hanging minus the handset. If in luck, the reassuring purr of a dial tone will have been heard, then details of the smoke will have been relayed in a nervous fashion to our Fire Control. Within minutes, this passer-by will have heard the braying of the two-tone horns as the two fire pumps raced nearer.

The fast-rising, heavy-looking smoke that indicates a hot, rapidly developing blaze is also beginning to bank down as the winter dampness exerts its downward pressure. The sodium lamps are being masked by this smoke pall as the two pumps squeal to a halt outside the burning tenement. On the opposite side of the street, a gaggle of 'night birds', on a hen night perhaps, and in between drinking spots, are flocking together, their necks craned sharply back and eyes drawn to the drama high above their pavement perch. Their white faces and curvaceous, exposed mini-skirted thighs turn blue, then white, as the spinning light of the beacons falls on and off them.

The nearside front door of the Pump Escape opens even before the vehicle has fully stopped. Out emerges the heavy bulk of Station Officer Ben Tuke, in charge of the duty watch this evening. His six-foot-plus almost allows his black-booted feet to touch the street from his seat. He dismounts and angles back his white-helmeted head sharply, as he takes in the belching hot smoke 50 feet above with a critical eye – an eye sharpened by attendance at scores of such scenes over his many operational years within inner London.

The tenement block to which we have been called has seen more than its share of fires over the years. It is divided into flats and a high proportion of its residents are men employed in manual work. The risk of fire in these sorts of premises is high, as within each flat there is a cooker, often a fire hazard. Most call-outs are to pans of food left on stoves that have overheated whilst the tired-out resident, fatigued after a long day of heavy labour, flops out in a chair or bed, forgetting the pan on the hot ring. Men have died in fires within this building, and who can be sure at this stage whether or not a similar fatal outcome might arise.

As in other parts of this locality, a fair proportion of the population is Irish and many of the residents within this tenement are of that stock. Men who, from the early 1950s onwards, had made the often rough sea passage to Holyhead, then on to the capital if not to seek their fortune then a solid regular wage from the work to be found there. The Irish male is famed for his ability to carry out often back-breaking manual work with relative ease; work that many others could not handle.

Life for these men was often hard and austere. Many of them were single, remaining in that non-marital state all of their lives. Arising at some ungodly hour from a lonely bed, then making sandwiches and a flask of hot drink before going to wait for the construction company's wagon. After a day of hard graft all that awaits are the four dingy walls of a seedy, dimly lit room and tiny kitchenette.

It is little wonder that many hard-working men found solace in alcohol and comfort in the company of those whose lives were following the same rugged road. But if alcohol possesses the power to temporarily transport a man onto a happier plain, it can also hasten that same man's demise.

For these men, once the pubs closed, it was the custom to head for a popular venue such as the Galtymore Ballroom in Cricklewood Broadway, the Forum in Kentish Town or the Gresham Ballroom at the Archway end of the Holloway Road. Within these so-called 'ballrooms of romance', amidst music that told of such places as Galway, Mayo, Sligo, Connemara and the hills of old Tyrone, hard-working folk could find temporary refuge from the weekday drudge and be taken back across the sea to Ireland.

But of course there are other ways of gaining this release without venturing out every weekend to pub or ballroom. Some might choose to stay within their bare rooms. Alone or with a friend, they could slip away from their confines within the heady vapour of a few bottles of Irish whiskey. Enveloped in the sharp smoke of many cigarettes, they could yarn about their youth in the old country or listen on a record player to such well-known songs as 'Galway Shawl', 'The Fields of Athenry', 'All for Me Grog' and 'Whiskey in the Jar'.

Sub Officer Dolan Devine, himself a Mayo man, had experienced the hard and lonely domestic existence of many of his countrymen over 20 years ago. Then he too had arrived in London as a single man and had lived in the kind of down-at-heel accommodation to which we have responded.The incidents of fire attended in such premises since he had enrolled with the brigade almost two decades earlier have been numerous.

Another such incident is now unfolding before his eyes as he drops down from the appliance and trots over to Ben Tuke to confirm what his guvnor's operational plan will be. The heavy smoke issuing high above the street is no different from that encountered at hundreds of fires attended. However, at this late hour in a tenement with such a poor safety record, it is highly likely that many occupants will be asleep, oblivious to their surroundings. Most fire deaths occur during the sleeping hours and the greatest threat is that created by smoke. In this case, it is possible that someone has already been overcome by the deadly toxic gases within the blaze. It is such knowledge that causes that lubricant of urgent action called adrenalin to course through the veins.

Ben Tuke's crews, like those of Biff Sands, are effective and dedicated men. Their abilities have been sharply honed by the high number of emergency calls received within some of the brigade's highest fire-risk districts. As such, our actions are virtually automatic. Even after being woken by the call bells from a deep slumber, our operational inputs with hose and ladders are of a very high standard. They are a testimony to the experience and wisdom of those old school guvnors who insist on daily drill yard practice so that the public who help pay our wages receive a superior service when emergencies arise.

Even before Dolan Devine has conferred with Ben Tuke, men are throwing open lockers, locating with standpipe in hand the yellow plates that indicate a fire hydrant, grabbing a long line or pulling off yards of hose-reel tubing with an adrenalin-fuelled swiftness.

John Lydus, 23 years old and with a superior level of personal fitness resulting from cross-country running, has yanked the hose-reel drum furiously, spinning it at such a speed that a good 100 feet of the three-quarter-inch rubber tubing is already on the street ready to be taken up to the blaze on the sixth floor.

'Dolan, mate. Get yourself around to the rear of the block and check out the scene as quick as you like,' shouts Ben Tuke.

Leading Fireman Dick Friedland and 15-year veteran Dennis Breedon have already rigged in BA.

'Let's get a crack on, Dennis, mate. I've been to a couple of nasty jobs in this effing place and this one looks as though it's going to be a right bit of "thickers",' shouts Dick, using the old London Fire Brigade term to describe a fire that is generating extremely dense smoke.

Firemen Rusty Checkland and Dick Brizer are already inside the block, anxious to get up to the fire floor. The resilience and fitness of youth is aiding their run up the mosaic-tiled stone staircase, that is until Ben Tuke's bulky form enters the ground-floor hallway. 'Hey, you two! Bloody well slow down! Stop running and pace yourselves,' he barks at the two youngsters, who have less than three years' service between them. 'If you keep running like that, you'll be breathing so hard by the time you are in the smoke you'll be gasping in a lot of toxic shit and finish up being overcome. Then you'll be no use to any bugger, yeah? So pace yourself, lads,' he says firmly, but his voice is tinged with an air of benevolence. For he remembers how green he had once been.

When Ben Tuke was a new recruit, he was schooled by the old firedogs, their fire-fighting experience etched in the facial creases around their eyes – wrinkled hallmarks on skin overly dried by the sting of acrid smoke, an occupational hazard of all firemen before the use of breathing apparatus became mandatory. In his youth, there were few, if any, who would have dared argue about the time-honoured practice of 'getting in' to a fire. It was the fireman's unwritten creed to save lives, protect property and carry out any other humanitarian duties without question.

On this night I am detailed to ride the Pump and within seconds of John having pulled off that huge loop of hose reel, I have grabbed its pistol-like nozzle. Together we drag the tubing into the entrance. I'm pleased to see that the internal stairs will facilitate a quick hauling of the hose by line up the well, rather than having to drag it around corners, which wastes a lot of time and requires a lot more length.

Even though it is now well past midnight, the staircase and landings on the lower floors still reek of the cooked cabbage of earlier meals. But at the fifth floor this is replaced by the pungent odour of burning bedding and wood. It is a scent familiar to firefighters everywhere – an acrid smell; a nose-puckering smell that hints of burning autumn leaves mingled with the scorching of an iron left for too long on its rest.

The smell of smoke is a useful indicator as to what is likely to be involved in a fire before its seat is reached. In the small hours of the night, it can send out a powerful signal and a reminder that the situation might be one of great urgency.

The smoke at the sixth floor is now heavy, almost viscous and gravy-like in appearance. Its colour is a mid-brown with intertwined yellow strands, which indicates the presence of sulphur, a product of incomplete combustion, and a warning that a flashover, a sudden explosion of flame in which all the combustibles in a room ignite, could be a potential hazard.

The heat within this malevolent pall has pushed the smoke up to the ceiling from where it mushrooms down, giving off a scorching, radiated heat within its deadly breath. Although the gaps around the windows of the burning rooms have allowed a venting release for some of the smoke and heat, we must still quickly locate anyone trapped and find the fire's seat. We must extinguish it before the whole floor bursts into that fireball of flame, as the great heat raises the temperature of all the combustibles to the point where they will ignite explosively.

Ducking very low under this baking heat cloud, and with the hose reel and its nozzle looped around my arm, I prepare to crawl towards the room in which the fire's source is suspected. John is right behind me, hauling the tubing, which is in turn being fed by Dick Brizer and Rusty Checkland from the landing, where conditions are more tenable.

As we crawl along the greasy, worn linoleum towards the apartment, the heat increases and that thin layer of breathable air an inch or so above the floor thins. I begin to cough as the acrid fumes seep into my nose and mouth. I momentarily stop crawling, flatten my face to the filthy floor and feel again that life-preserving cooler vein of oxygen. I draw this in through my nose, forcing my mouth to keep shut – no easy task when the tension and exertion cause you to gasp.

After a few seconds, the coughing slows and we press on, pulling ourselves along half on elbows and hands. I can see above the almost totally obscured light from a bare pendant bulb. It must somehow be sheltered from the thick rolling veil for me to detect it, but see it I do. Below its dim illumination, I can just make out the hunched form of a man, coughing loudly. He is on his knees, directly outside a door off the corridor along which we have crawled.

'My mate Seamus . . . he's still in there, you know. Aye he's still in there,' he coughs out, with an Irish accent as thick as the smoke all around us but in the slurred, barely coherent speech of a man who sounds worse for wear from drink. 'Holy Mother of Mary, I had to leave him. The smoke was too thick, it was too hot. It's awful. Can you get to him, please?'

We all know that Ben Tuke is a smoke eater from the old school and he demonstrates this in no uncertain fashion. Here we are, flat on the floor, grovelling around in that barely breathable channel of air, when the footfall of his size 13 fire boots announces his presence. He is upright, his head in the thickest, hottest layers of this evil smoke, and he isn't even coughing as he drops down onto his knees at my side.

'You're doing fine, son, fine. But there's no time to waste. I've got a couple of BA men on their way up, but you must make a search of the far room as fast as you like, if any fucker is going to come out of this job in one piece.'

'Got it, guv,' I splutter, as he grabs the inebriated Irish man's arm in his thick fingers and leads him to the relative safety of the landing.

We continue, turning right, crawling over a threadbare excuse for a carpet, trying as hard as we can to keep mouths closed to shut out the vile and hot toxic smoke, praying that our nostrils will filter out a lot of the poison rather than our succumbing to its lethal ingredients before we have located the victim. Even as I crawl I am wondering how blokes like Ben Tuke can eat smoke like he does; whatever the reasons, I know this smoke is punishing us harshly and behind me John is coughing uncontrollably. It feels as if we have been inside this inferno for ages, when in fact it has only been a few minutes since we first started our rescue and fire location efforts.

Encouraged by the sure knowledge that the two BA men will soon be with us, and spurred on by the information that there is a person trapped, we press on. I can now see a dancing ruddy glow about two feet above the floor some fifteen feet away, the red reflecting off the brown and yellow smoke. As we get closer, I can just make out what looks like the rectangular lump of a double bed. It is from the top side of this that the flames are leaping. Their curling tongues have reached up to the ceiling and spread laterally to each side, forming a moving ruddy T-shape of smoke-masked flame. I unloop the hose reel from my arm. The pressure built up by the water behind the pistol nozzle can be sensed like a coiled spring awaiting its release. I squeeze the trigger and the hose bucks a bit, like the recoil of a high-powered revolver, as I give the flames on the bed a good burst. I sweep the jet around and this produces a hissing heat of steam, which momentarily darkens the glow only for it to light up almost at once.

'Take the reel, John. I'm gonna start a search. We ain't any time. It's too bleeding hot in here. Cool the room down, mate,' I cough out as I pass the nozzle to him.

He opens up the spray and whirls it round and round in a desperate effort to lower the room's temperature as I begin my search for Seamus, who we suspect will be on or around the bed.

My hands feel under the bed but meet no human form, only the old-fashioned chamber pot, the wet obscene contents of which splash onto my hands.

'He's on top of the bed – I can just make him out,' coughs John. 'He's there, on your right. Let's grab him and get out of this hellhole before it goes. It's too effing hot. My ears are tingling!' he blurts out.

As I get to my feet, I also sense the blistering heat on my ears, a feeling like a hundred hot needles within this superheated brown woollen shawl of smoke and deadly gases.

Sub Officer Devine, here in this tawdry tenement amidst his native folk, appears after checking that no one is hanging out of windows at the rear of the premises. Now he is alongside us, squatting on his haunches in this furnace-like heat. Dolan is another man capable of eating smoke, but had it been food he was consuming, the question would have been: 'Where does he put it all?' There is nothing of him. Small in stature and small in facial features he is, but his heart and spirit when involved in fire and rescue are as big as anyone else I ever knew. There is precious little Dolan has not experienced. His 'at the sharp end' nose for danger sniffs out the perilous conditions surrounding us all. He is holding a small axe and with a few deft strikes takes out the upper panes of the windows overlooking the street from which was first seen the percolating smoke pall. This lifesaving piece of 'good hand' firemanship vents the pressurised heat and smoke, releasing it to race with great heat-driven velocity up into the north London night, thereby lowering the potential for a deadly flashover.

As Dolan is doing this, my hands fall onto a stubble-covered head, the skin of which is slimy with heat-induced perspiration. I move my hands down over a barrel chest, bulging belly and thickset thighs, which have literally been cooked by the burning bedding. John gives the bed a good 'drink' from the tubing, knocking down most of the visible flame and creating a cloud of dense hot steam.

John then takes hold of the legs, whilst I get my arms around the broad chest and back, the thick flesh disguising the tough muscles and sinews beneath, the product of a lifetime handling a shovel, spade, sledgehammer and pick. Seamus is, as they say back in his native land, 'a real broth of a boy'. What would he have given now, I think briefly, to exchange the heat of this virtual funeral pyre for the cool rain and wind blowing on a Connemara shore?

BA men Dick Friedland and Dennis Breedon are with us now, the click-clack of the valves on their sets signalling their presence. 'Take over the reel, we might as well get him out now,' coughs John to the pair as we half drag, half carry the very heavy form back to the relative coolness of the landing. Here the hands of Rusty and Dave reach out to help us lower Seamus onto the cool tiled floor.

Dolan is squatting in the corridor, looking grey and shattered from his sterling work in venting the deadly smoke and heat. Next to him our Station Officer is still erect but in a corridor now almost free of smoke. The weak yellow light of that bare bulb suspended from the old brown twisted flex casts its jaundiced light on the still steaming Seamus. Rising with that steam is a sickly sweet smell of burned human flesh. The whole of the Irish man's back and thighs are severely burned, so much so that the deepest fibres are visible; he looks like a charred version of the human musculature printed in bright colours on an anatomical chart.

'The poor sod. Get a burns sheet up here, pronto,' instructs Ben Tuke to no one in particular. Within minutes, young Rusty, barely out of his probationary year and experiencing his first nasty job, has skipped down the stairs and reappears with a burns sheet from the Pump. This is gently placed over the charred limbs in an effort to lessen infection within those horrific burns.

Ben Tuke knows that with that degree and depth of burning it is unlikely that Seamus will survive. He had once read in a medical journal that an individual's survival prospects could be estimated by adding the percentage of the body burned to the victim's age and if it neared 100 per cent chances were remote. If the victim is the 50 or 60 years he looks, then the 50-plus percentage of burns makes things look very bleak. We just had to hope that medical advancements had increased the prospects of such a grievously burned victim surviving.

In any event, it isn't the fire services' job to adopt a negative approach and miracles have occurred before. We never give in. Those decisions are the lot of the medical profession.

If proof were ever needed of the determination of the fire services' front line personnel, one has only to observe Dick Brizer, young in service and years, on his knees, bent over Seamus, into whose lungs he is forcing his young, life-sustaining breath. While there is life, there is always hope and, as I look down at this scene, my own head now starting to clear, I observe the slow rise and fall of the victim's chest.

Into my mind comes the inscription that is on the awards made by the Royal Humane Society, imploring rescuers not to give up: *Peradventure a little spark may yet lay hid.* Young Dave continues to search for that spark until relieved by the ambulance crew a few moments later.

As the fire is under control and to get fire cover back on the local stations should other emergencies arise, the Pump Escape is put back on the run, after which both crews begin the process of ensuring that the fire is fully out and that no extension has taken place into roof voids or adjacent rooms. Whilst we are doing this, Ben Tuke begins his investigation as to the likely cause of the blaze, including requesting the brigade photographer, who will record the fire scene in case the victim dies, which will result in an inquest and the need for a coroner's report.

The previously dense smoke has now all but cleared within the affected rooms, yet the fire has again left its pungent calling card to show that it has inflicted its damage to life as well as to cloth, timber, plastic, brick and glass. The two rooms and kitchenette, squalid and unsavoury before the fire, are certainly in a mess now. The walls and ceilings have been blackened by the greasy smoke. The burning bedding and mattress had ignited an adjacent foam-filled sofa. This, in its turn, had set fire to an old wooden sideboard, the varnished veneer of which added its fuel to the hot fog of fire. The double bed is now a blackened, burned, water-sodden mass. To the right of the windows which Dolan Devine axed to vent the great heat, there is an old oak wardrobe, superficially charred and with one door ajar, revealing a scruffy old gabardine raincoat.

'Seen this, guv?' I query, in between pulling down the heat crazed and damaged plaster to check for still ignited embers. I point to the right-hand large open pocket of the raincoat where the top half of a partially consumed bottle of Irish whiskey is visible. A few moments earlier I had noticed amidst the wreckage of the room two other full-sized whiskey bottles, now heavily smoke-stained but with their labels still readable. Both of them are empty and I point them out to Ben Tuke.

'Looks like they were having a right bleeding beano, don't it?' he remarks with a wry grin.

'Sure does, guv. I guess the guy you helped out will have a right tale to tell, eh?' I reply. Although the man we found outside the affected rooms will likely be in shock, it is vital in pinpointing the cause of the blaze for the Station Officer to speak to him whilst events are still fresh in his mind – or as fresh as they can be, if the empty whiskey bottles are any indication of the amount of alcohol consumed.

Ben Tuke was a first-class mentor and he respected my enthusiasm and keenness to learn, so I was delighted to hear him shout, 'Get someone else to do the damping down, son. I want you to bring your notebook and pencil and come with me. You can do the scribing that will help with the fire report, yeah?'

So our big guvnor, still tall even with his large combed white helmet under his arm, and with me tagging on behind, locates the room where a resident has said the man will be. He knocks on the paint-peeled brown door, the steel number plate of which has fallen loose but left its imprint below. After a couple of minutes and the sound of shuffling from inside, the door is opened by a large, heavily built man with a red moonface and with a bald pate covered by some combed-over hair strands. He has big outward-turning ears and hands the size of small shovels. He is wearing a dark green woollen cardigan over a grubby white shirt. On the collar are some small specks of dried blood, probably the result of a cut whilst shaving because two pieces of tissue paper are stuck to his bull neck next to the inside of the shirt collar. His trousers are brown corduroy, shiny with wear, and the laces on his boots are unfastened and trail on the floor.

'Good morning, sir, sorry to bother you at this hour, but I need to have a word with Seamus's friend, who I am told might be here. My name is Ben Tuke. I'm the officer in charge and it's important I see him soon,' he says quietly.

'Oh, yes. Do come in, sir. That will be Brendan you will be wanting, will it not? He's just through here. It has been a terrible night, for sure, especially with Seamus getting caught an' all that,' the moon-faced man, named Joe, goes on as he leads the even larger bulk of Ben Tuke down a dingy corridor. We are taken through a tiny kitchenette, in which sits a small square table covered with a stained green-checked tablecloth. On top of this is a small clear glass basin in which there are slices of yellow cheese; next to this is a large basin heaped with cornflakes. In the corner, there is one of those tiny cookers known as a Baby Belling. It has two small rings atop which is a blackened stew pot and a small red kettle with a blackened spout.

There's a small casement window overlooking a rear courtyard and below this a brown butler sink with a draining board. On it sits a scratched Oxo tin with a thick elastic around it, no doubt a workman's sandwich container.

Even before the room at the far end is entered, the sound of loud snoring can be heard. On entering, Ben Tuke can see in the dimly lit room a white-haired middle-aged to elderly man. He is sprawled unceremoniously face up on a faded old red sofa, fully clothed save for his boots, which sit on the floor beside him. His mouth is half open and the snoring is that of a man who is sleeping off a skinful of strong drink.

'Brendan! Brendan, you gotta wake up, yer know. It's the officer man from the fire brigade, who needs to talk with you,' Joe shouts as he grasps the man's shoulders in his huge ham hands, shaking him until he starts to stir.

Brendan stares and blinks, trying to gather his fogged and fuddled senses and to come round from the grogginess of an alcohol-induced sleep.

'Now, sir, do you and your young man here fancy a mug of hot tea to be supping while Brendan gathers his wits? I guess that you will be parched after being in all that smoke and heat, will you not?' enquires Joe.

'We could kill one,' replies Ben Tuke, 'but tell me, do you know Seamus and Brendan well?'

'Oh, yes. I have known the pair of 'em for well over 25 years. We came over here at the same time for the work on the construction sites and we've stuck together down the years, for sure we have. I'm a Tyrone man myself, whilst Brendan here is from Mayo and Seamus comes from a little town called Glenamaddy in Galway, where the showbands used to play back in the '60s,' he rattles on in that fast delivery peculiar to men of his land, as he pours the boiling hot water onto the tea in a large chipped brown pot. 'We all met up on the ferry to Holyhead. The work back then was too good to miss, as there was little chance for good money back home at that time, for sure. None of us ever got wed and we used to earn very decent wages, which allowed us to send a fair bit back home each month to Mam and Dad down the years. Sadly, they have both passed on now,' he says with a sad sigh as he points a thick rod of a finger at a card on the wall on which there is a faded photograph of an elderly couple. It is a remembrance funeral card familiar to those of the Catholic faith. 'Still, when the big man above calls, we all have to go at some stage,' he continues in a subdued voice as he pours the steaming tea into four large pint pot mugs that have blue rings on their outsides and are chipped around the edges. Then, as though the recall of his parents has reminded him of mortality, he looks straight at Ben Tuke and, in a hushed voice, asks, 'Is our dear friend Seamus going to be all right, sir, if I can ask that of you?'

Ben looks at the man's red face for a good few seconds before replying, 'You must remember I'm a fireman not a doctor, but in over 25 years in London I have seen people as badly burned as Seamus survive, but they were I think a fair bit younger. So it is not easy to answer, but he will receive the best attention at the Burns Unit in Middlesex and we can only hope and pray for him.'

'It looks as if it is in God's hands now, for sure,' responds moon-faced Joe and, as he finishes that sentence, Brendan staggers into the kitchen.

'What's in God's hands, Joe?' he queries as he enters the kitchenette, his snow-white hair dishevelled and his clothing crumpled.

'Ah, there you are, Brendan. The fire brigade officer here needs to have a word about the fire and we were talking about poor Seamus's burns an' all that.'

Ben Tuke stands up, his tall bulky frame dwarfing the squat shortness of Brendan, and he holds out his hand. 'I am sorry about your friend, but I need to try to find out how the fire started.'

'How's me old friend Seamus? Is he badly hurt, sir? Where have they taken him?' he asks, his speech now barely slurred as the effects of the alcohol recede and the traumatic nature of events begin its sobering process.

'Here, get this down you, Brendan, and why don't we go in the back, so that you can tell the fireman what happened,' says Joe.

Brendan takes the mug of tea and together we all go into the bedsit. Over the next half an hour, with the room lit by a dim low-wattage bulb and the red glow from the hissing gas honeycomb wall heater, Brendan gives his account of what happened and how it had ended in such dire consequences for his friend of over two decades. I frantically make notes of the salient facts.

It turned out that because Seamus had been feeling very tired after a day digging a trench, a task which didn't get any easier when a man was nearer 60 than 50, they had chosen to spend the evening in the tenement. They had decided to splash out on three bottles of Irish whiskey and with several packets of cigarettes sit down and reminisce about their youth and the old country and play some of their favourite tunes and songs on the old portable player. Brendan said that by about nine they had seen off the first bottle of whiskey, which they were taking with water, and that they were enjoying themselves with the heat from the gas fire and the smoke from the cigarettes making things seem very pleasant.

Because of his tiredness Seamus had begun to succumb to the drink ahead of Brendan. As a result, he had got onto the bed to lie back, with Brendan sitting on the old settee next to him. Seamus had dozed off and was snoring. The whiskey bottle was on a small table next to the settee. Alongside it was an identical empty bottle, which was filled with water for diluting the whiskey 'to make it go longer', as Brendan described it. He recalled having another half tumbler of whiskey and water before dozing off himself, as the fug of heat from the fire and the drink took its effect. He had woken up with a sharp smell in his nostrils.

Looking towards the bed, he had seen a blue haze of smoke and realised that Seamus must have fallen asleep with his cigarette still burning and that the top bedding must have been set alight. Brendan had leapt up and shaken Seamus, using some of the water in the bottle to douse the sheet.

Brendan refilled the bottle with water and opened another bottle of whiskey, and soon they were imbibing again, enjoying the filter-tipped cigarettes. He recalled to the best of his memory that they must have got through a good half of the bottle and that this had taken them up to about a quarter to midnight, a time when normally they would be leaving the pubs to go to the dance halls. Brendan soon felt his eyes closing, he said, as 'The old drink had really begun to speak to me, for sure.'

A strong urge to relieve himself had brought him back to consciousness, if not to a clear mind, but almost at once on regaining his senses, he smelled that pungent odour of smoke. He opened his bleary eyes but could barely see a thing. He rubbed his eyes furiously, thinking that his eyelids were stuck together. Then it dawned on him. The lack of vision was due to thick smoke in the room.

Once again, Seamus's lit cigarette must have set fire to the bed. Indeed, through the now thickening pall of smoke, he saw red and orange flames on the bed leaping upwards, with his friend still in the middle of it all. For a second or so, he had frozen. Then in a panic he had screamed to Seamus to wake up. He grabbed the whiskey bottle nearest himself from the small table, the one containing the water, and poured its contents on the flames. There had been a 'vump' of a noise and a blue flash and the red flames had worsened. To his horror, Brendan, in his alcohol befuddled panic, had picked up the bottle half full of flammable spirit instead of the one with the water in it.

In spite of his inebriated state, he had the presence of mind to get to the kitchen and fill a pan with water. But in the now dense smoke, he could not locate the bedroom. The room was full of hot, eye-smarting, lung-irritating fumes. The bed was invisible, with Seamus unconscious somewhere on its now rapidly flaming form.

Station Officer Ben Tuke listened without interrupting. Occasionally, as he listened, Brendan wiped a tear from his blue eyes with a strong, stubby hand, calloused and roughened by years of gripping the shafts of spades, shovels, sledgehammers, picks and drills.

What a terrible tragedy this had been. Two old pals, once young men with all the sweet anticipation of youth. Young men with a sense of thrill and anticipation arising from that great adventure in crossing the Irish Sea like so many of their forebears, fuelled by the excitement of moving to a city where it was once believed the streets were paved with gold. Then realising down the years that things are seldom what they seem. If never finding their financial fortunes, however, they found a solid friendship that would stand the hard test of time.

What a terrible irony that the alcohol that had so often eased the often rough and rugged road of their existence had played its part in ending the life of Seamus, who sadly succumbed to his terrible burns a few days later. Another victim of the many fatal fires that were attended during that period, and a death that would so sadly sever a bond of friendship first forged between three men on that passage across the Irish Sea over twenty years earlier.

Chapter 4

London Bound

The 1970s were a decade that saw real changes in attitudes, fashions, fads and philosophies, many of which had taken root in earlier times. It was the era of such TV series as *The Sweeney*, which, along with others, seemed to herald the demise of a more traditional bureaucratic or starched-collar way of operating. *Dixon of Dock Green* was being rapidly replaced by the gung-ho attitudes prevalent in characters like Regan and Carter, and other mavericks such as Clint Eastwood's Dirty Harry. And London was at the heart of the revolution.

Looking back, I feel sure that the increasingly permissive, anti-authority opinions of the '60s and '70s, as illustrated by those TV cops, played their part in diluting the traditionally raw strategies and policies of fire and rescue that would herald a new era. This renaissance saw improved communications and brought about progressive technological change, along with the arrival of new legislation on health and safety, which was both a blessing and a curse.

This shift from the strictly regimented, militaristic philosophy of wartime meant that the 1970s probably was the last decade in which most supervisory officers, including the higher 'brass', had heavy practical experience and were steely authoritarian characters – not dissimilar to the much respected and feared hospital sisters and matrons who had worked their way up to positions of authority via a hard practical route from the ground floor upwards.

Given that many London Fire Brigade officers, both of junior and senior rank, had seen either full-time or conscripted military service, or had fought the conflagrations caused by the Blitz, such tough 'in your face' attitudes were not surprising.

The terms 'political correctness', 'sexual discrimination', 'equal opportunities' and 'health and safety' had still to enter the lexicon of the fire service; when they did, the fire service was forced to change in ways few who were serving during the 1970s could have envisaged. Perhaps it was the age of innocence. It was certainly the age of raw courage and unquestioning discipline, both of which saw many lives and properties saved.

So it was against this background that, within a few days of spying that *Sun* photograph, my application to the London Fire Brigade was in the post and I was seeking a transfer in my rank of Leading Fireman. At the time, the capital's fire force was experiencing some problems with recruitment. Not only was the pay not attractive enough to encourage applicants to put their lives on the line – you could earn more in the safety of a factory or office – but potential applicants already holding rank outside the capital were put off by the high cost of accommodation, especially if only one half of a marriage was bringing home a wage. The London weighting allowance did little to encourage applications either, it being a meagre amount compared to the financial burden of city life.

Within a week, I received a letter saying that at that time there were no vacancies for Leading Fireman, but vacancies did exist for Sub Officer, the next rank above mine, and Firemen. Because I had 'tickets' of qualification up to Station Officer, the third promotion rank, I could have applied for a Sub Officer post and secured it, but my wish to gain a solid practical experience, which included learning how best to maintain, as far as possible, the safety of any crews that I might one day be responsible for, set off warning bells about running before I could walk. This wasn't Wakefield or Castleford. Even Leeds, with its thousand trades and population of over half a million, could not approach the fire and life risk of the capital. There was probably as much life risk in one

East or West End inner-city street as there was in twenty streets in most provincial towns, with the exception of places such as Birmingham, Manchester and Liverpool. Then there were the subterranean risks posed by the vast underground rail system, plus the high number of mainline rail stations, Heathrow Airport and the docklands, which, although virtually dead, was still at risk from arson-related fires.

The other thing I had to remember was that even though you might be a junior officer on a quiet suburban station, any major incident calling for many appliances and crews could see you in the thick of it.

No, I had my theoretical qualifications, and they were very instrumental in my early years in keeping me safer than I might otherwise have been, but if there were no vacancies in my rank, then there was only one thing to do and that was to revert back to Fireman. I was still only 25, time was on my side and I had to remember that I was a northern incomer, far removed in birthplace from the sound of Bow Bells.

I had soon secured a bedsit in Fulham, which would suffice during the month I would spend alone doing my transfer course at the brigade's training centre in Southwark before my wife Carmel joined me.

There were only four on my transfer and re-enrolment course at Southwark and the three others were former London firemen who had left to chance their arm at other occupations. They were all cockneys. Two of the men were in their 20s – one, as I recall, had been a stoker on the old steam engines before joining up for a few years, while another had emigrated to South Africa, but it hadn't worked out. The thing I remember about all of them, though, was that they had young children to feed and this came out in the intensive month-long course – I sensed that they were desperate to get through.

In terms of getting up to par with the drills and equipment, it was harder for me than them, I felt. The first thing was that after nearly a year as a junior officer, partly responsible for a watch, here I was back at the start again. I had known this, of course, when I made what some thought was a crazy decision to revert and give up that one silver shoulder bar, but I did wonder in the first week if what I had done was correct. I had let my usual decent level of physical fitness slide, what with the house sale and upheaval of such a move. Unlike the other three, who were much more au fait with the London equipment, it was mainly new to me.

The one-ton fifty-foot extended wooden-wheeled escape ladder, still the principal rescue ladder back then, was a heavy beast to move around and to manually wind up the extensions. I remember getting quite a slagging from one of the others during one drill when my inputs didn't help the drill we were doing. Looking back, I reckon he thought that if my input made them all look poor he might not get back into the brigade, so with hindsight it was understandable.

The biggest hurdle for me was mastering the hook ladder. This 13-foot-long ash wood ladder was synonymous with the capital, even though other provincial brigades used to carry them too. My fellow recruits, although not having touched the ladder for a few years, did have the advantage of having done three months' training on them during their former recruit training. Such was the quality of that training, and so important then was the ladder to the brigade's potential to save lives in narrow alleys where no other ladder could reach, that if a recruit failed to master it, he would never pass out of training school and his career in London would be at an end before it had begun.

On the top end of the ladder there was a curved bill and back from that a row of serrated steel teeth. These teeth were there to grip into a wooden windowsill, and the bill hooked inside the window frame. We had to complete individual and paired ascents of the seven-storey, seventy-foot-high drill tower by hooking the bill over the first sill, climbing up, straddling the sill, pulling up and pushing up until the next sill and so on. In the paired drill, one man wore a huge leather belt to which was attached a large steel dog clip or climber's carabiner. The belt man snapped the clip into a stainless steel ring at the head of the ladder and then, being held only by the clip, had to pivot round until he was at a right angle, his inside leg braced tight and the other braced over the hard yard below. Both arms were outstretched and using a second ladder, the second man then pushed the ladder up to him awaiting secured only by the serrated teeth, the bill and the clip and ring. Once he had the second ladder, he punched it up to the sill above, hooked it on and then had to engage in the hairiest bit. This was the transfer from his own ladder to the second, now hanging from the sill above. If he got it wrong, because there were no safety lines or nets, it might be, to quote that famous Bolton steeplejack Fred Dibnah, 'a half day out with the undertaker'.

My only experience five years earlier with the hook ladder had been during my recruit training when we ascended the training school drill tower individually, so I had never seen a hook belt nor done the two-man drill before. With the encouragement of a good instructor, plus the other three trainees, I managed to reach a passable standard on a piece of equipment that most within the Fire Brigade's Union wanted to see rendered obsolete. To those who considered themselves dedicated firemen, the hook ladder was not only a great confidence builder – if you could trust to its sway, every other ladder seemed like a staircase – but also a badge of excellence, and a ladder that set the London Fire Brigade apart from its provincial cousins who, in most cases, had stopped drilling or even carrying it several years earlier.

London is a unique city and, like Paris, where the hook ladder originated, has lots of narrow alleys and closed courtyards inaccessible to any other ladder. The dedicated fireman believed that even if it was only used once in ten years to save a life that otherwise would have been lost, then it should remain a part of the rescue equipment. There must have been literally thousands of ascents made in station drills since the ladder's introduction in the early 1900s until its withdrawal in 1984. Many lives were saved by it and only a few lives were lost during training in 80 years.

So I mastered the hook ladder and gained a refreshment on the Proto BA set, which I had previously qualified in back in my own recruit training in Leeds. After completing the course, I spent six months serving at such West London stations as Acton, Fulham and Hammersmith before I took the plunge and got a transfer to the very busy East End. By that time, my wife and I were living in a rented flat in the pleasant north London suburb of Muswell Hill, just below the tall BBC TV mast atop Alexandra Palace, or 'Ally Pally' to the locals.

Before my transfer south, quite a few of my workmates had warned me of the character of the London fireman. One calculated put-down that cropped up often was calling them 'big jet men' – which implied they did more damage with the water they streamed into a building than the actual fire had done itself. 'They have so many pumps and such good water supplies that they wash burning buildings down the road,' they would laugh. I found this amusing, especially when I remembered one incident while serving on a West Yorkshire station where a TV and radio dealers shop was 'flooded' as a consequence of a large diameter hose being directed through a first-floor window into the smoke punching out. A close reconnaissance revealed that the thick smoke was from a pile of burning cardboard boxes that a small hose reel soon extinguished!

Opinion on Londoners seemed to have arisen from a variety of sources. There were the old army sweat Yorkies who had been billeted with the Private Walker-type spiv of *Dad's Army* fame during wartime or national service. Then there were the junior officers who had attended courses at the Fire Service College at Moreton-in-Marsh and who remarked on London firemen as 'real gobby bastards', 'trappy buggers' or 'crafty cockneys'.

I think another reason behind the negative reactions came from visits to the capital, where it was hard for some to get their heads around the size of the buildings and the proliferation of street fire hydrants, some being 24 inches in diameter. When your Yorkshire patch includes a few farms where the only water has to be relay pumped a mile or so, or the street hydrant might be only six inches or less, then the idea of the 'big jet men' becomes more understandable, albeit an erroneous one in my long experience.

Any fears I had about being an incomer never materialised. I had to endure the expected 'E by gums' and 'There's trouble at t' mill again', but nothing more. I am sure that if a cockney with a pronounced London twang had transferred to Yorkshire during the 1960s and 1970s, he would not have been so readily and quickly accepted as I appeared to have been. And I don't think that it was just good fortune. London, like New York, is a very cosmopolitan city and has for many years been a melting pot. Within the East End in particular, the immigrant population comprises Dutch, German, Russian, Irish, Italian and Caribbean communities, with an influx from the late 1960s from the Indian subcontinent. The surnames of London Fire Brigade personnel when I transferred illustrated some widely dispersed geographical origins. It was because of this that I seemed to be more readily accepted. After all, in a brigade where soft Scottish burrs mingled with hard Glaswegian, and Irish from the north and south mixed with lilting Geordie and machine-gun Scouse, why should a Yorkshire accent be so much at variance with them?

I had, of course, transferred to the inner city for the action to be found on the front line of this huge urban conurbation, but I never envisaged how much of life and raw humanity I would be witness to by day and night and of how tremendously addictive was the sense of anticipation: the never knowing at what second those call-out bells would toll their urgent clamour, never knowing what we would find on arrival.

And, of course, never being sure that we would return unscathed.

Chapter 5

Death and Sweet Sherry

In the 1970s, those who had travelled along life's highway for 70 or 80 years by then would have been born in vastly different times. An 80 year old in 1972 would have first seen the light of day in 1892, with his own parents probably born on average 20 to 30 years before that. Likewise many of the huge houses previously occupied by the upper classes, more than a few of whom would have been prosperous enough to have employed a cook and a few servants, remained as clear reminders of the thick lines of demarcation that had existed between the haves and the have-nots in London, as elsewhere in the country. After the Second World War, and the Conservative-led plan to invite Jamaican citizens over to the UK, to fill vacant posts within the transport and nursing sectors specifically, these houses and the occupancy of them began to witness real change. A property developer or landlord with a good business sense could make money through rental income and premium deposits for door keys, but with a minimum input on their part in terms of property maintenance, and so began an exploitative market in rental accommodation.

Such was the wicked exploitation of either immigrants with poor English, the frail, the mentally weak or the elderly that the term Rachmanism came into the language, following the conviction of the greedy, unscrupulous landlord Peter Rachman, who terrorised tenants so as to obtain premium rent from the next unsuspecting client desperate for accommodation. *10 Rillington Place*, starring Richard Attenborough, referred to a North Kensington street frequented by Afro-Caribbean immigrants and a cross-section of these desperate tenants at the time, showing what these once proud family homes had become: a mix of lodging house, with cramped and damp garret rooms in which the potential for fire to strike was great, especially in the winter months (when use of paraffin heaters could be lethal as a means of staving off the bone-chilling cold).

Similar types of properties existed at most points of the capital's compass in the deprived and down-at-heel districts. A glance at the incident logs at North Kensington's Faraday Road station or Holloway's station in Mayton Street off the long and busy Seven Sisters Road would show that during a period from the 1950s up to well into the 1980s many fires were attended in these once proud houses that had been converted into smaller apartments or had become what are termed 'houses in multiple occupation', where a common kitchen would service many individual rooms.

It was in such a house – once a bustling family home for the upper classes but now a squalid run-down district – that we were called in a case of death and sweet sherry.

As is normal for me when I'm on nights, I sleep poorly, so even though it is well past midnight I am sitting in the mess room doing a crossword when the call bells clamour. In a second, I am dropping down the pole and in the rear cab of the Pump. My two crewmates arrive a few seconds later, their eyes puffed up and their faces a deathly pallor with the rude, heart-racing awakening. It's just before 3 a.m.

As Station Officer Biff Sands pulls himself into the officer's seat at the front, he shouts to the driver over the growl of the diesel engine that it's a fire call. He gives him the address, which is one I recognise, having attended several nasty house fires with fatalities in the surrounding streets.

'Be ready to get rigged!' Biff shouts, knowing that at this hour most folk are asleep, oblivious to fire, which means we might be dealing with thick smoke. Getting rigged means donning BA, which takes time. If someone is trapped, an entry will be attempted by non-BA crew, for sure, but if men are already rigged and just need to stick the BA mouthpiece in without fastening the body straps, all the better. You risk getting those straps snagged under a door or piece of furniture, but rather that than lose vital seconds that can mean the difference between a victim living or dying.

There is only a small amount of smoke visible under the street lamps as we pull up at the address. It is coming from around the edge of a window in a basement flat that faces the street. The flat is just one of several in this Victorian mid-terrace three-storey house. Other than a couple of cats the street is deserted, but like most inner London streets it is nose to tail with parked vehicles and this can impede our rescue operations if ladders are required to be pitched to rescue persons on upper floors.

Biff Sands has already descended the stone steps to the basement's front window and I follow him, after pulling off about 50 feet of hose-reel tubing, just in case it is needed.

I put the back of my bare hand onto the window glazing. It feels quite hot, and I can just make out the orange flicker of flame through what is a veil of smoke inside the room, looking like a yellowy-brown net curtain.

Breaking a pane of glass to release the build-up of smoke inside a burning room can save a life. But it can also kill someone who might have lived if it is done at the wrong time, which is why there is an art to ventilating smoke and heat in structural fires.

If the heat in the building has already risen to the stage where the combustibles present are ready to ignite, an injudicious shattering of glazing or opening of a door can introduce the extra oxygen that the fire needs. On more than one occasion, well-meaning members of the public and police officers have forced a door or window in a rescue attempt, the result of which has been a bursting fireball of flame, meaning the end for anyone still inside and danger to fire crews outside. Good firemen never ventilate without having a charged hose line ready, hence the hose reel I had pulled off.

After his 20 years on some of the brigade's busiest stations, Biff Sands can read fires better than a copy editor can read text. He knows that the window glazing heat and the smoke that is visible is not at that highly dangerous stage yet. So when he instructs me to ventilate, I don't hesitate to reach up and tap out a small portion of the uppermost part of glass of the sash window. Immediately, the brown smoke inside pours out. At this stage, we know that conditions inside will rapidly improve for whoever is in there, but we don't know how long the fire has been smouldering away for and it might already be too late.

Biff has already forced the flimsy lock on the front door and within thirty seconds two men without BA have entered the flat and located an unconscious victim atop a bed. They carry her out and I can see that it is an old woman.

Her skin is hot and she is slimy with perspiration; the same perspiration, caused by the fire's heat, has stuck her thin but long hair to her scalp. I check her mouth for obstructions such as dentures and can see none in the yellow light of the street lamp. I get a whiff of a sweet odour, which might be alcohol, as I put my black silk neckerchief over her mouth as a germ filter, tilt her head back and pinch the nostrils before blowing into her mouth. I watch her old chest rise and notice the white fuzz of facial hair on her heavily wrinkled face.

There is no pulse in the carotid artery in her scrawny neck and Dick Friedland, our Leading Fireman, commences cardiac massage in between my mouth-to-mouth inputs of breath.

A priority message ordering an ambulance has been sent from Fire Control, having received the emergency information, and its horns can be heard in the distance. Even at this early hour, London streets contain traffic and, to us battling to save a life, those horns represent hope. Within another minute, the ambulance arrives and its crew takes over and continues resuscitation before loading the woman on board and speeding her to the hospital a few miles away.

Once she has gone, I go into the basement flat. There is an ashtray on a small bedside cabinet that is stuffed with dog ends, indicating a heavy smoker. The fire had involved the bedding, possibly started by a still ignited cigarette that the old woman had dropped into a fold in the sheets as she fell asleep.

If a cigarette end still burning is trapped in a fold of sheets or the upholstery of combustible material, the temperature can slowly rise and lead if not to flame then to deadly smouldering smoke.

There is a half-drunk bottle of sweet sherry on the same bedside cabinet and an empty large sherry glass. Under the bed there are at least a dozen empty sherry bottles. I had detected what I suspected was alcohol when working on her outside and I can only conclude that here was an old woman who was either an alcoholic or a lonely individual for whom the drink was a great comfort and aid to sleep in the long and lonely watch of the night.

The accommodation is a tiny 12 x 12-foot room. It has a tiny tabletop oven and a separate two-ring gas stove next to a small sink and drainer. A wall cupboard and a tiny fridge contain some basic items of canned food; the remnants of a half pint bottle of milk and a half-used sliced loaf of bread stand on a small table. Other than the now fire-blackened and sodden single bed positioned alongside one wall, the only other furniture is a single wardrobe and an old tallboy. A small television set occupies a table next to the front window, which we broke to ventilate the smoke and heat.

On the wall above the bed, there is a handsome framed certificate with a black frame. Although the smoke has covered the glass with a thin greasy coating, I can still make out what the certificate says. It reveals that the woman in the room, whose name has been obtained from one of the other tenants, is the same name as on the certificate, announcing that she had once been a member of the Royal College of Physicians.

She had qualified as a medical doctor in the 1920s, meaning she would have seen more than her share of death and disease, and deserved better than to see out her final years in such surroundings.

A month later the coroner recorded a verdict of accidental death as a consequence of inhaling fire fumes caused by a still ignited cigarette setting fire to bedding. He added that the alcoholic content of her body indicated that she was probably intoxicated before she inhaled the deadly fumes. She would have understood the medical findings far better than those of us who had tried to save her.

Her passing was a poignant reminder of the irony of life. It told a story of two noble but so differing professions; of an intelligent, ambitious young woman who had spent long years at medical school at a time when academic rigour was so high, a woman who might then have spent years as a hospital junior doctor before entering into whatever branch of medicine she chose, in an era when bias against female doctors still existed.

Our calling, with its primary objective of saving life, was no less noble than hers, but the highly academic entry requirements to qualify for medical school, then to become a physician, contrasted with the minimal academic qualifications required to become a fireman.

The antiseptic aura surrounding medicine contrasted strongly with the hot, smoky, grimy arena of the sparsely furnished room in which her life had ended so tragically and alone.

Chapter 6

Crewmates and Characters

'So, how are you finding life as a fireman in the Big Smoke? Must be a huge change from dealing with haystacks and cattle stuck in mud up there in the frozen north,' says Paddy Mulligan, as we sit enjoying a cup of tea during our evening shift stand-down.

'Not a lot different, Paddy, from you rescuing leprechauns and heifers from Irish bogs before you took the old ferry to Holyhead, eh?' I respond with a grin.

Paddy is a short, stocky man in his 40s with tousled black hair going grey at the temples and twinkling blue eyes. He had been a part-time fireman in Northern Ireland before following many of his countrymen during the 1950s over the water to seek more lucrative work in England.

He had laboured on building sites, delivered beer and worked the door at inner London nightclubs before enrolling with the brigade over 15 years ago. Strong as the proverbial ox, he was a good man to be working with; you knew that if you got injured, he would be the first to drag you out. He had a cousin who had crossed the wider water of the Atlantic to settle in New York and he was also a fireman with New York's Bravest, as the members of the New York City Fire Department are known. I had visited the Big Apple myself and stayed at a couple of its busiest fire houses and knew that Paddy would have fitted in there just as well as he did here.

'Well, now, since the height limit was lowered, I reckon there are more wee leprechauns in the brigade than there are back in the old country, for sure, my old son,' Paddy responds with a twinkle in his eye.

'Come on, Paddy. Next thing you'll be trying to tell us that we'll soon see women serving on the front line,' says Leading Fireman Dick Friedland, one of the elder statesmen of our watch, who is looking towards his retirement in 18 months' time after over 30 years' service. Fifty-four year old Dick, originally from Israel, had witnessed some truly horrific scenes during his military service when he helped liberate many of his fellow Jews from the Nazi death camps. Like many of his peers, he has noticed the slow changes that had taken place in so many aspects of life since that time. The war seemed to him to have been the catalyst for the ending of the old order of things and over the years he has seen little to convince him that standards and attitudes are not changing for the worse.

Paddy, the combative Irishman, is always ready for a lively argument and cannot resist responding to Dick's comment about females on the front line. It wasn't until 1982 that the first female firefighter joined the London Fire Brigade.

'And why not, for sure, could we not have those of the fair sex serving alongside us? I would give up my dormitory bed for any of 'em!'

'Yeah, sure you would, as long as you could then get back in it with her,' I quip.

'To be sure, you have a fair point there,' Paddy replies with a chuckle. 'But to be serious for a moment, don't women serve in the front line of the army in your ancestral homeland, Dick?'

'Well, Paddy, I will grant you that's true up to a point, but the culture there is totally different and I can't ever see it being agreed that in this country women will be seen on the front line of fire and rescue,' Dick retorts.

'Don't see how you can think that, Dick,' Sub Officer Jack Hobbes butts in, as he rolls a cigarette between his nicotine-yellowed fingers.

Jack had been with the brigade for just under 20 years. He was a no-nonsense Glaswegian who had served a few years in the busiest stations of the West End within the division we all knew as the 'Royal A'. He's 41, about 5 ft 9 in., with a fine head of steel-grey hair and a trim grey moustache.

'There have been women coppers for years. In the early days, it's true, they were only deployed on cases involving children and women, but today they do just the same job as blokes,' he goes on.

Jack has a point, I muse, while sipping my hot mug of tea. The subject of women working on the front line of our job was not something much aired. This was mainly because few people, men or women, believed that in a world where little boys wanted one day to become a 'fireman', it would ever be possible for females to become a front line 'firewoman'.

Women had worked in the service before, during and after the war. During the war, some had driven the vans and cars that pulled trailer pumps, but the majority worked in control room or administrative posts. Their performance was exemplary under great stress, although not on the operational firefighting arena, and there were few who believed that at a future date they would be given the opportunity to take their place on the front line. Then again, I continued to muse, no doubt way back in the 1920s before the first metropolitan policewomen were recruited into the force, policemen had then also refused to believe that women would one day work side by side with them.

When I enrolled, the height limit was 5 ft 7 in. and a would-be recruit had to have a minimum unexpanded chest of 36 inches, with at least two inches of expansion. Although the chest measurement remained, in 1967 the national appointment regulations saw the height lowered to 5 ft 6 in. It had been rumoured that this reduction was due to a shortage of recruits willing to enter our potentially hazardous occupation and, of course, as I've mentioned already, the relatively poor pay, especially in the capital, where there were much better paid and far safer occupations to be had, did not help. The thinking behind knocking an inch off the height requirements was that the door would then be opened for the short but strong guy, such as the builder's labourer, who, in many cases, was stronger than some of those who met the height and chest standards.

But central government had been making noises for a while about bringing in legislation related to discrimination on grounds of sex or race and some new 'equality of opportunity' laws were imminent. You didn't have to be Einstein to realise that major shifts, such as a reduction in physical standards, might be seen as the thin end of a wedge that had the potential to seriously undermine the service's operational competencies.

By the same token, the EC-driven proposed health and safety at work legislation from which the fire service would not be exempted had a similar demoralising effect. When such radical proposals were combined with the changes likely as a consequence of laws relating to sexual discrimination and equality of opportunity, it was not difficult to perceive that some very heavy reverberations could arise. These might be so strong that the fire service's traditional role and its operational effectiveness could be severely compromised. There were more than a few who feared that health and safety law in particular could become a strong steel shackle that would inhibit the fireman's natural and developed instincts for protecting those in peril.

When I think back to the fine characters I worked alongside, I remember that I learned early on not to judge a book by its cover. There are more than a few people in life who draw all too quick conclusions about a person on account of their appearance, their stature or their accent. No doubt there were those who on hearing mine had at once thought of the stereotypical northerner. Perhaps to them I came from a land of textile mills and tall smoking chimneys, or black canals and back-to-back *Coronation Street* houses, all mixed with some magnificent countryside full of sheep, populated by characters from Lowry paintings, or men with cheap 'muffler' cravats and a whippet dog, or a loft full of racing pigeons.

By the same token, it was not too hard on hearing 'Cor blimey' or 'All right, mate' to be reminded of the stereotypical cockney scene and its characters, the kind who populated *Billy Cotton and his Bandshow* of the 1950s and 1960s, or of the old-time music hall stars, some with a cigarette in hand, the ash about to fall onto a waistcoat stained with the marks of cheap good living. 'Cheerful

Chappie' Max Miller, although from Brighton, was to everyone like the typical cockney, with the rapid patter and a quick mind.

Perhaps one of the younger men who put me in mind of him was Ricky Tewin. Here was a man who typified that inner London world seen by many from the north as comprising pearly kings and queens, pie and mash and liquor and jellied eels, not to mention the good old knees-up sing-alongs around a sawdust-floored pub's honky tonk piano.

Ricky had, thanks to a family connection, gone straight from school into Smithfield Meat Market in Clerkenwell, scene of a tragic and extended subterranean fire back in 1958. Hearing his strong accent brought to my mind all of those images, especially when he would often suddenly sing out a few lines relating to the Aldgate area of the East End where he was born and lived. The accent did not speak of culture or a bright brain, but I discovered that you cannot easily judge a man or woman by how they sound. In spite of the hard manual labour of his job, carrying half of a pig's carcass on his back, white smock and hat stained with blood, here was a man with a quick mind and one honed by night school classes, where he had studied politics and sociology. If you add to that a big social conscience, you have the ingredients for some lively late night mess room debates involving this man in his late 20s, with his average build, which disguised a strength built up by his physical work before joining the brigade several years before my move.

His foil, if you like, was fellow Yorkshireman Barry Priestley, a former face worker at a coalmine not over-far from Doncaster – that is to say, a man who actually hewed the black gold from low seams. In his mid-30s when I first met him, Barry was a stocky man of average height with an upper body physique that would not have looked out of place in a professional wrestling ring. Barry was the man to have in a crew when strength was needed. Again, to hear him speak with the hard consonants of South Yorkshire, you would be forgiven for thinking that he was all brawn but few brains. You would have been so wrong, as he could offer considered opinions on all but the deepest of subjects and there were not too many of those heard in fire stations, whether in the capital or in the provinces.

When your career ambitions and progressions take you among several fire stations, as mine did, then you meet and work with a wide range of character types, often with very varied backgrounds. During my years in the inner parts of the capital, I must have worked with several scores of firemen and junior officers and, with very rare exceptions, they were all dedicated and competent souls. Some had the sort of personality and spirit that make you remember them better than others. Those names that feature most prominently in my accounts appear because of the impression they made upon me, though that is not to say an occasional mention should be taken to mean that those fellow crewmates were any less competent or dedicated.

Many of my mentors and the crews I worked alongside had experience as either regular or conscripted members of the armed forces. More than a few had seen action during the Second World War, or in Korea, Malaya, Cyprus or Northern Ireland. Tall or short, stocky or slim, these fine characters brought a solid wisdom and great experience of life and humanity to the brigade. As the decade unfolded and these characters retired or stepped up the ladder away from firefighting, they took their noble qualities with them; their militaristic approach to fire and rescue work slowly transforming in their wake. As a reflection of society in general in that era, the principles upon which their philosophy was based eroded with their passing.

To my way of thinking back in the '70s, and I still hold these views today, the greatest sensation of change must have been perceived by the Station Officers and Assistant Divisional Officers (ADO – the next rank up from Station Officer) in particular, because these were men who, in being usually a fair bit older, had been around longer and would have had more solid benchmarks by which they could monitor a changing society.

The London fireman would use the term 'guvnor' or 'guv' when addressing all but the higher ranked officers and there were some blokes who wouldn't flinch at using that term to address the Chief Fire Officer. The term at its simplest indicated who was the boss: it was the standard terminology for the Station Officer who managed a watch or an ADO who worked from Divisional HQ, and would take command once six pumps had been assigned to a specific incident.

There was and is a definite 'them and us' attitude in the fire service, as in the armed forces. Back in the '60s and '70s, there was a greater proportion of men who were much more deferential to higher ranks than would be the case in more recent years. No doubt this respect had been grounded in seeing military action or by being involved in the Blitz raids on London, when it was imperative to know that a crew would operate obediently and react instantly to any safety command given.

Many of the men I worked alongside in London did not hand out plaudits easily and they could smell an incompetent boss a mile away; within minutes, such an officer's reputation would have spread to stations at the other side of the division. There is no doubt in my mind that it was the massive experience and sure-footedness of the best guvnors that had given the London Fire Brigade its excellent reputation for shirking from nothing in the protection of both life, primarily, and property.

Given that these fine leaders feature so prominently, something of their pedigree needs to be explained, because when I arrived most of the guvnors whose wing I was under had been born back in the 1920s and 1930s, an era where youngsters were much more deferential and respectful to their elders.

There were still officers around with a similar temperament and attitude up to about the late 1980s, and there are some around even today who would fit well into that mould, but the forces of political correctness and the tight shackles of health and safety law, plus a sea change from the 'in your face' to a more 'softly softly' management approach have effectively outlawed the direct kind of management commonplace to those guvnors featured in this book. To better understand what made these characters tick, we need to understand the historic forces which drove them.

It is so easy to forget that although the normal human lifespan is not that long, a person may have witnessed much social change within his or her lifetime; the mistake is often perceiving 'history' as having happened long ago, when in fact this is not always the case. For example, when I became a recruit in the late 1960s, it was easy to forget that it was not much over two decades since the 1939–45 war had ended. Within the London Fire Brigade during the early 1970s, when I first arrived there, a fireman who had completed 35 years' service and was about to retire at the compulsory age of 55 would have become a recruit in about 1937.

A fireman about to retire in 1937 after thirty years would have enrolled only seven years into the twentieth century and been born towards the end of the nineteenth. So, even in the 1970s, there were men within the London Fire Brigade who had worked with men who had enrolled in the early part of the century.

It follows that transferred attitudes stemming from Victorian times towards discipline, respect for those of a higher rank and a belief in the ethos of a public life and property-saving service were ingrained. This was especially the case in our essentially conservative society, with its clear social divisions and in which there was natural deference to those higher up the social or employment scale.

On top of this, it has to be remembered that the UK fire service in general, and the capital's fire brigade in particular, was rooted in military procedures and practices. Many of the earlier London firemen had been mariners. These were favoured because they had demonstrated during their long voyages that they could withstand the tedium of inaction without revolt, making them capable of withstanding the periods of inaction between emergencies. London, for all its buildings and populace, was nothing like as operationally busy then as it would come to be in the ensuing years.

Sailors used to climbing masts on sailing ships were, of course, not afraid of heights and, with a proven temperament towards obeying instructions, they were ideal recruits. Even as the century progressed and 'old salts' formed a declining percentage of recruits, the two world wars and other theatres of conflict across the globe in which British armed forces were involved provided a cadre of men in possession of similar qualities to those of their mariner forebears.

Some members of the London brigade, already in post before

the outbreak of war in 1939, signed up with the armed forces. Others, some of whom had previous military experience, remained in post. The courage and dedication of those serving in the capital during the war has been well documented, but their exploits, and the lessons they learned about the vital need for discipline and order during times of peril, became solid foundations to the way in which those leaders managed crews in future years. Such strong foundations were bolstered by the tough attitudes of those men who had seen active front-line military service on land, sea or in the air.

I will not forget the character of some of the instructors at the brigade's Southwark training centre who had seen such war service.

The level of safety held by fire crews depends on a number of factors and one of the most important of these is the quality of the training imparted on the recruit course, because this lays the foundation for the whole career.

It was clear that the passionate way in which these men instructed their charges with the most safe and effective methods to use when handling hose, holding branch pipes, climbing ladders of all kinds and wearing BA sets came from an inner vision that drove their lecture deliveries. There was an urgency about the information that was so skilfully delivered that would be seldom seen in instructors in later years.

When I later learned of the theatres of war in which such men had served, I began to better understand their philosophies. Some had been aircrew in damaged aircraft that had limped home from bombing missions, their safe return being largely down to the hard discipline of their 'skipper'. Others had seen action in the Burmese jungles or in the brutal fighting in the Korean conflict of the 1950s. These good men, like so many of their wartime peers, had stared death and their maker in the face and lived. I am convinced that one cannot go through such harrowing situations without having your appreciation of the gift of life being made that much stronger.

When they gave lectures or supervised practical drills, they seemed to bring to them that sense of discipline and that appreciation of how an undisciplined, sloppy approach can mean the difference between coming out of a dangerous job in one piece or being carried out in a body bag.

Station Officer Ben Tuke was a swarthy man of over six feet and of solid build. With sparkling eyes and a lantern jaw that gave him a no-nonsense, severe appearance, he could bring a rowdy crew back into line by a mere stare. He would never pull rank and the day officers of his era and philosophy had to resort to invoking the national disciplinary regulations (replaced by industrial tribunals in 2005) to manage his troops was the day he would retire.

Ben had joined the Infantry at 17 and did several years before joining the brigade in the 1950s. He had served for a long time on some of the capital's busiest stations, including the high-life-risk hotel and boarding house districts of the Bayswater area. He had been commended on more than one occasion for his valour.

Ben was a 'smoke eater' from the old school and I have described his seeming resistance to thick smoke within Chapter 3: Whiskey in the Jar. Such had been his experiences deep within the bowels of smoke-logged premises or in crawling along heat-and smoke-filled corridors and stairways that he knew better than most when things were becoming 'too hot'. He drew heavily on his extensive experiences when supervising our BA training. This was carried out either in a purpose-built smoke chamber at a central or divisional location or back at the station using such places as a basement in pitch-black darkness to simulate dense smoke. He would have us crawling across this basement, searching for a dummy he had previously jammed into some remote corner. If your knees started to hurt and you tried to take the weight off them by raising an inch, you would feel the heavy crack of a 2 lb torch across your helmet to the bellow, 'Keep low if you don't want to burn yer ears!' – the forceful blow simulating the skin-peeling heat contained within the superheated smoke, below which we were forced to crawl.

On other occasions, our daytime cook would jump back in shock from the sink where she was peeling the spuds as we entered the window head first from a ladder pitched from the High Street far below.

Ben Tuke could have been Biff's twin in appearance, attitude and experience. He was hewn out of a mould of deep and wide practical experience and later forged on the hot anvil of many, many fires in all types of premises.

The practical drills such guvnors employed were grounded in their numerous experiences in those countless incidents. To men like them, doing drills that were unrealistic did little to prepare men for the frenetic and rapidly changing scenarios that can exist at serious fires. But as health and safety legislation progressed, these real-life scenarios were outlawed and more than a few would come to believe that we were worse off for these restrictions.

Biff Sands was another huge guy who had joined the army as a teenager in the 1950s. He had seen service in Korea and admitted that the sight of seeing mates killed because they had become complacent and did not believe what their commanding officers had warned played a huge part in the way he managed his firemen. He often used to say that you're just as dead if caught in a flashover or by a collapsed wall as you are from a sniper's bullet.

The armed forces had taught men like Ben and Biff that constant, repetitive drills certainly brought about the ability for men to work almost on autopilot. An almost instantaneous reaction following an officer's instruction to, for example, slip and pitch the wheeled escape could mean those vital seconds between death or survival.

Not all of the much-respected guvnors of those years were big and hard looking. Joe Kinneal, a Northern Irishman, had barely reached the then minimum height limit of 5 ft 7 in. and weighed no more than ten stone. He had a serene pale face with pink cheeks and twinkling blue eyes under a shock of white wavy hair. Truth be known, he looked more like a priest, and in fact he was known as 'Father Joe'. However, he could also be extremely firm when on serious emergencies and, again, like the others described, he had won his spurs the hard way, by numerous responses to all manner of emergencies.

Perhaps Joe's greatest claim to fame was that he had served in the East End for the whole of the war and had attended most of the conflagrations in the docklands created by the Blitz raids and the carnage caused by the V2 rockets towards the war's end. He had, at the time I worked alongside him in the mid-'70s, spent over 35 years in and around such East End stations as Whitechapel, Shadwell and Shoreditch.

Stepping up a rank, ADO Tom Monsal, featured within several of my accounts, was an extremely cool and competent officer. Firemen on the extremely busy East End stations did not hand out plaudits easily, especially to those who were senior to their own much respected watch guv, but Tom Monsal's operational competence was almost legendary across the East End divisional stations. He had spent most of his 30-plus years within the East End and had been a guvnor on the busiest stations for many years before receiving the rank markings of ADO.

When we were on a nasty job with real danger all around, it was comforting to see Tom Monsal drive up to take command. No chance of him placing crews in needless danger by opting for the kind of ill-thought-out strategies made by the few with a scant operational background who had somehow slipped through the filtering net of heavy experience.

He had a developed instinct and sound judgement forged by his having responded to countless emergencies He had seen it all: severe fires in tenement blocks with occupants screaming for help high above the street during the early hours; warehouse infernos in the docks, with the ever-present threat of floor and wall collapse onto crews below; fires in heavily smoke-logged deep basements and sub-basements, in which it was so easy to become disorientated and lost. Then there were the non-fire emergencies involving road traffic or the rail network, both above and below the streets. Nothing made Tom Monsal lose his composure. He was truly a 'good hand' officer.

The respect that men had for him meant that if they heard 'Tom Monsal's coming on,' they knew that if he couldn't help them bring a job to its best and safest conclusion, no one could.

So the big strength of these officers, and the one major element in their make-up that gained them such respect and loyalty from their men, was their massive operational experience.

All of these Station Officers, along with people like Sub Officers

Jack Hobbes and Dolan Devine, and Leading Fireman Dick Friedland, and all the others with whom I served, were from a mould that rarely existed in later years.

When I joined the service in the late '60s, it was unthinkable for these fire-forged officers not to make a determined effort to 'get into' buildings and to stay there until the fire source was located and extinguished. Although the saving of life was always the first priority, we still had to extinguish fires when no one was trapped; after all, that is what we were paid to do, without which businesses would be lost and people put out of work. If a highly experienced senior officer arrived at a blaze and found crews directing a jet from the street when it was possible to get into the building and suppress the flames from within, the crew would feel that officer's wrath, with such shouts of 'What yer doin' – washing effing windows?'

But slowly the 1974 Health and Safety at Work Act (from which the UK fire service was not, like the armed forces, exempt) began to bite and attitudes began to change. As the decade progressed, a minority began to question if we were paid enough to risk life and limb by going into premises on fire in which no persons were reported trapped. Of course, there are occasions when people are still inside a burning building, even if those outside say otherwise.

In later years, when I had become a guvnor myself, I well remember such a situation in a fire involving a fast-food restaurant where the duty manager was insistent that all of his staff were accounted for. How wrong he was when BA crews heard the sobs of a young woman trapped by choking thick smoke inside a basement store room. Without our customary precautionary search, and the bravery of a man who risked his own life by placing his face mask onto the teenage girl, she would have perished.

I have no doubts that any practical competencies that I held owed a tremendous amount to those cool and steadfast officers who mentored me whilst still a callow young man from Yorkshire's broad acres.

So, although there were those who started to move away from the raw but heroic and noble actions of earlier years, most of the guvnors I worked alongside did not fall into that category. It was significant that the greater their experience on the front line, the more firm was their view that without that aggressive initial push, an already hazardous situation could worsen, placing lives and property at further risk.

These then were the guvnors during the 1970s. More than a few of my colleagues of those busy years found it a worrying reflection of a much-changed societal attitude when, in later years, where style seemed to be more important than substance, form more important than function, such fine and courageous characters attracted such labels as 'dinosaur', 'diehard', even 'dangerous' from some quarters. But tellingly, and more often than not, these comments came from those who preferred the sinecure of the relatively low risk and quiet outer suburbs on which not all crew were as dedicated to their work as others.

Any fair-minded and keen fireman who worked out of the capital's busiest stations in those years would have found such descriptions ignorant and unfair in the extreme. They had a supreme confidence in the leadership qualities of their guvnors.

To them you only enrolled with a uniformed lifesaving emergency organisation if you were fully prepared and willing to give a dedicated service. That dedication required you, like the crew of lifeboats, to put consideration for those in peril before your own safety, so as to preserve life whenever it was humanly possible.

They say that old soldiers never die but only fade away. Sadly, many of those mentioned have passed on, but what has not faded is the memory of all that these consummate public servants stood for.

Those memories and their indomitable spirit burn as bright today in my mind as those raging infernos and killing flames to which we so rapidly responded side by side by night and day within the Big Smoke.

Chapter 7

Death from on High

'Fire on eighth floor of high rise. One person jumped.'

The imprint of a hand is clearly visible on the inside pane of a soot-blackened kitchen window – eight floors and eighty feet above the frontage of the high-rise tower block where the lifeless body of a young man is now lying spreadeagled across the hard concrete. A pool of blood is congealing around his head and his lifeless, open eyes stare up to the darkening sky.

We have been called at a little after eight on this late summer evening. A thin grey ribbon of smoke can be seen rising from the top floor as we near this large council-owned estate. As the two appliances halt in front of the block, the man's body can be seen. It is about ten or so feet in front of the building. It is clear to even the newest recruit that for this unfortunate person life ended instantly on the cold, hard pavement.

The address to which we have been called is one to which we turn out on a regular basis, though those responses are normally to rescue persons shut in the lift or to fires in the stinking refuse chutes. Not a tragedy the like of which has occurred tonight.

Sub Officer Dolan Devine cranes his head back until the rear of his helmet touches his back, looking up to the smoke cloud that is issuing from a top-floor window directly above where we are standing.

'Charge the riser and get the high-rise gear,' instructs Dolan.

The soft Irish brogue is still as strong as it was when he enrolled with the brigade many years before, following his employment as a cable layer with a well-known local Irish company.

We rush to lockers and take out the hose, branch pipes (a technical term for the hose nozzle), lines (used for lowering casualties/bodies and hauling equipment) and the BA sets, which brigade instructions mandate we use at fires in high-rise dwellings. As we do this, Johnny Carre, the driver of the Pump, begins to connect the hose into the external coupling of the 'riser' (the short term for a rising main, either wet, i.e. charged, or dry, which has to have water pumped up it by a brigade Pump). This allows us to pump water up the vertical pipe to a valve situated on each landing. Niall Pointer has already entered the block. He has activated the 'Fireman's Switch', which allows the brigade to commandeer the lifts and speed up our progress to establish a bridgehead on the floor immediately below the fire floor.

'Effing hell, Sub. The lift's knackered,' shouts out Niall to Dolan as he enters the ground-floor lobby.

'Then for sure there's nothing we can do but walk up, is there now?' responds Dolan in his unflappable manner.

'Send an informative: *Smoke issuing on eighth floor. Lift inoperative. Crews making way to fire floor. Further messages may be delayed. One male apparently jumped from block to ground level – apparently dead. Police attendance required*,' he instructs to Niall, who sends it at once from the Pump. This radio message will inform senior officers of the situation and assist in their decision to attend.

Our fire boots, four pairs in unison, echo loudly, along with our laboured breathing, as we lug the high-rise gear up some seven flights of hard concrete steps and along landings – steps, landings and past walls that are daubed with lurid graffiti scrawls. The stairs reek with the stench of urine and are littered with cigarette butts, spent matches, empty fag packets and the odd contraceptive sheath, the grey contents glinting under the weak white lights on the cold brick walls.

Although now five floor and fifty feet above the ground, we can easily hear through the slatted timber ventilation ports the screaming whine of the fire pump, thrusting the water up the vertical pipe at sufficient pressure to overcome the frictional resistance of height and still give us enough hose pressure to fight whatever fire might be present. On top of this noise, however, another sound rises. It is the rushing, splashing 'shish' of surging water. As we arrive on the sixth-floor landing, we are met by a cascade of water racing down the stairs.

'What the hell! Some bastard's only had the landing valve away, I bet,' snarls Dolan, droplets of sweat from the effort of the laden climb running down the ruddy complexion of his face and dripping from his sharp nose. 'Jim, get yourself back down. Get John to knock off the water and then fetch up a couple of lengths of the small hose. We'll need to set into the sixth-floor valve. But for Christ's sake, be bloody sharp now. We don't know what we have here and the buggered lift has lost us minutes, to be sure, all right it has,' he blurts out between his panting breaths.

On each landing there is a brass valve opened and closed with a hand wheel. In this deprived and vandal-ridden neighbourhood, such an item will be like a glinting jewel to the magpie beaks of some residents, who could earn the price of a good few pints of ale from its exchange at the local scrap metal merchant.

Niall's exit at the ground coincides with the arrival of the third Pump. In minutes, the extra hose has been carried up to the sixth-floor landing. It is connected and by use of a hose and hand-controlled nozzle the water will now not reach the missing valve on the floor above.

'Where the fuck have you been?' a very irate and panicking man of about 25 screams. He's half-running, half-pacing back and forwards outside an eighth-floor flat. 'I fucking called 999 ages ago from a neighbour's phone. My flatmate's inside. Oh my God. I slipped out to the pub. When I got back, the frigging flat's ablaze. Will he be OK?' he rattles on in that sequence of rapid uncoordinated thought and speech that is the product of shock, fear and not knowing what has happened whilst he has been away.

Whatever fire there had been must have vented itself via the kitchen window because on entering the flat the atmosphere is warm but quite breathable without recourse to BA sets. In the centre of the living room, which has a connecting door to a small kitchen, there is the half-burned remains of an armchair, blackened and still smoking, in thin wisps. A small pile of what appears to be burned clothing is on the floor immediately in front of the chair. Other than the inevitable smoke and soot-stained walls and windows, the flat is undamaged.

'It looks to me as if there has been a slow build-up of smoke,' says ADO Tom Monsal. He had decided to come on from Divisional HQ after being informed by control of Sub Officer Devine's informative. 'Dolan, have you had a chance yet to speak to the deceased's flatmate to verify if anyone is a smoker? Looks to me as if the poor sod down on the ground might have fallen asleep in the armchair and somehow set light to a newspaper, and then the clothing or armchair covers,' muses Monsal.

Dolan is about to say that he will go and speak to the flatmate when an anguished cry comes up from outside the flat. A police officer has just told the man that the dead body lying outside is his flatmate, whom he thought was still inside the flat.

'Do you know if it is just the two blokes who occupy the flat, Dolan?' inquires Monsal, his hawk eyes settling on an open sliding-door wardrobe in the rear bedroom. In that wardrobe there are about 30 pairs of different coloured stilettos neatly lined up below rails of skirts and dresses.

'I'm not sure, Guv, at this point. But just before you got here I overheard a neighbour saying that a couple of hard-faced women looking like streetwalkers were always coming and going at night, along with two men in and out at other times.'

'Maybe the women and the men are the same people,' he suggests.

Tom Monsal's mind contained an extensive mental file of hundreds of fires attended during his long inner-city service. One image sprang to the forefront of his thoughts. He recalled another fire a few years back, an arson attack that had involved a couple of cross-dressers. His old boss had called it a 'transvestites' tiff' and he pondered now if this was a similar incident. Alternatively, the women's clothing and shoes were perhaps the property of local prostitutes, given the flat's location close to an infamous red-light district. It was common knowledge that in the world of vice violence against prostitutes by their pimps or by rival vice kings resentful that their own patch was being encroached upon was an occupational hazard. Arson fires as one form of retribution by these pimps was a possibility that couldn't be discounted.

Monsal then walks into the small kitchen and studies the hand imprint on the hinged opening section of the soot-covered window. Through the murky glass he can see far below the revolving blue beacons of the appliances and those of police vehicles. A doctor has certified the man as dead. He can see the scuttling black forms of police photographers and others now investigating why and how the young man has ended up on that unforgiving pavement.

After about ten minutes the ADO turns and asks Dolan to come into the kitchen. 'In my opinion, Dolan, I think the occupier was probably a smoker who had fallen asleep reading a newspaper. The smoking materials have somehow ignited the paper and the resulting flames have ignited the man's clothing and the combustible coverings on the chair. I took a look at the body before I came up and the front of his pullover and trousers appear burned and scorched. The heat and burning might have aroused him. Imagine you're in a heavy doze and wakened by burning. When you come to, the room is full of choking smoke, so you panic. You'd be totally shocked and disoriented. He's then instinctively gone into the kitchen for water or to try to get to the window for air. I guess he could have climbed up onto the sink units. In his confusion and panic, he climbs half out of the opened bottom bit of the window. He was only a skinny little sod. He could have slipped through and tried to grab the window frame, hence the slurred handprint.'

'It could well be that was the way it was,' replies Dolan.

'Whatever it is, Dolan, old son, it's essentially a police matter now. You know the drill. Don't let any of our big-footed fuckers disturb anything and you'll need to make a critical investigation of the fire cause, including getting the brigade photographer out, yeah?' continues the ADO, adding, 'And I'll need your coroner's statements and fire report before you get off.'

'Yes, OK, Guv. There's no rest for the wicked, is there now?'

'At least you're in a lot healthier state than that poor young fellow lying cold as a fish on that hard concrete below. I'm away now, but keep me posted on things as necessary. I'm keeping an open mind at this stage on this one, but I do recall a similar job a few years back. Take it easy, Sub, and I hope I don't see you again tonight,' he replies, his parting words said in the hope that they both have a quieter time in the ten hours of duty remaining.

Several months later at the inquest, which the coroner had previously adjourned to gather more information, an open verdict was recorded. If there were any sinister reasons as to why or how the death occurred, they never came to light. Perhaps the theory put forward by Tom Monsal wouldn't have been too far off the truth as to how a young exile from Scotland had met his end on that fateful night in north London, yet another unfortunate victim of the ravages of fire and smoke.

Chapter 8

Coping with the Grisly Side

Some Personal Reflections

Many of the incidents I attended as a fireman involved fatalities. One of the reasons I made notes after attending emergencies, especially those involving loss of life, was that I hoped writing about it would help me, when still a very young man, to better handle the grisly side of the job. When I made my notes nearly 40 years ago, the fire service was a close-lipped organisation that seemed to revel in its modesty. It was the same 'silence' of our fathers and grandfathers, who preferred to bottle up their own gruesome memories of war's atrocities, which they had witnessed as young men. Back then, we lived in a virtually closed society, with no hint of the developments that would ensue in succeeding decades within the realm of communication.

I always understood that men who had suffered torture in an Asian prisoner of war camp, for example, might want to keep the lid screwed tightly down on their awful sufferings, but I often found puzzling the general unwillingness of so many not so affected to reveal details of what must have been seminal points in an otherwise ordinary life. This same understatement and modesty permeated much of the fire service.

I had long held the belief, however, that society would change and people would want to know much more about how jobs like the police, fire service and medical profession operated, warts and all. As I spoke more and more to members of the public, this became apparent and I knew from their questions that they were curious specifically about how fire and rescue crews cope with the grisly business of death and in our case severe injury of the victims of fire and non-fire incidents. After all, the public do contribute to our wages by their rates and have a right to know.

Scarcely a day passes now without fictional or documentary series and films being aired trading in raw, often gratuitous violence. The internet has enabled users to access all manner of bizarre scenes and because of this daily exposure the public's sensibilities have toughened. Perhaps if these latter-day wonders of electronic communication had existed in the first two or three decades post-war, detailed descriptions of what fire and rescue crews had had to deal with would not have been guarded like state secrets.

However, it is also pertinent that 30 or 40 years ago there was not the same availability of professional psychological counselling services as there is today if a man or woman is unduly troubled by what he or she sees.

It would seem common sense to ensure that all potential recruits to the fire and rescue service are required to undergo a psychological assessment. This provides an early indication of those individuals who might find attendance at gruesome incidents difficult to cope with. But even as I write this, my mind goes back to what many of my mentors in my early days within the capital would have thought. Back then it was a matter of personal pride to show your older, more seasoned colleagues that you could cut the mustard when on the more unpleasant incidents. One either got on with the job without whining or left for another occupation. And most found that they were eventually able to tolerate far more than they had ever envisaged – a healthy human mind within a balanced personality can cope with far more than you would ever imagine. In my experience, the more traumatic the scene that's witnessed, the greater is your ability to handle the gory and grisly side of the work.

And like it or not, the general mood amongst many of the crews in my day, including senior officers, was that coping with the grisly side of the work proved you were up to the task. If you couldn't cope without going weak at the knees, the inference was that the fire service was not the job for you.

That was one of the reasons that recruits were on probation – unless you have previous experience, how do you know if you can handle the sight not only of dead people but also the horrific injuries that sometimes accompany death?

Then again, the kind of gruesome incident that can test a character doesn't occur every day, even in the busiest stations. In the vast majority of cases in this country, on a four watch (shift) system, even in the centre of the largest cities, it is possible to go a very long time without seeing a dead body either in a fire or other incident. So a recruit could be well out of the probationary period without being aware if he or she could cope.

When I started making my incident notes, there were at that time far more of what we term 'working jobs' (serious emergencies and/or rescues) than is the case today and any man who wanted front-line action was free to apply to serve on the busiest units. It is perhaps a measure of the keenness of what can be termed the dedicated fireman that some of the busiest stations had waiting lists a mile long.

It was also on the busiest stations that you had the best chance of learning if you had what it took to become a competent 'good hand' member of the team. That competence was measured by your peers and by those who supervised you when the chips were really down. They would be keeping a watchful eye to see if you had the stomach and mental fortitude to cope with the scenes of blood and gore, plus the coolness to overcome the quite natural fear of the unknown when going into situations from which others are fleeing in sheer terror.

Personally, I found being a member of a disciplined, semi-military uniformed service somehow allowed me to numb my sensitivities at the worst 'dirty' jobs we attended. When fire and rescue staff are working in uniform as a team, they gain a strength from their shared sense of responsibility. This *esprit de corps* is not there when off duty and alone – I remember this well from that first incident when I came face-to-face with death, having come across the car crash en route home from work in Yorkshire.

When on duty, certainly at the outset of one's fire and rescue career, it becomes a matter of pride not to let your more experienced crewmates down by showing signs that you can't cope with trauma and death.

The 'stiff upper lip' so typical of the British must have played its part in the careers of countless employees faced with the raw side of life. The young airmen on bombing raids over Germany, the Spitfire pilots engaged in dogfights over Kent, the soldier or sailor seeing action on the front line of battle, the recruit ambulance attendant, student nurse or doctor. They all had to call on a controlled persona to cope with the reality of sudden death, especially when the ending of a life is through horrific injury. This control of emotions will always be necessary, for what use would anyone be in that line of work if they broke down in a sobbing heap rather than stiffened their nerves and got on with the task in hand?

Of course, no matter how well developed is an individual's sense of compassionate humanity, dealing with the deaths of people who are strangers will always be easier than dealing with the death of someone who is close. The psychological distance that exists between the emergency service front-line employee and the 'unknown' casualty does, out of necessity, provide a buffer that lessens the impact on the nervous system. With time, and the experience of attending more and more traumatic scenes, a mental callus seems to grow that desensitises the rescue service operative from the grim reality of a situation.

This having been said, there were some incidents that I attended that had a much greater impact upon me than others. One of the most significant of these was the Moorgate underground disaster of 28 February 1975. I was at the scene for the whole two days and two nights of my duty tour. My notes, made at the time, record how it was that the great familiarity of everyone on that horrific job with the underground as regular passengers made the rescue operations much more poignant than if the incident had been in an unfamiliar location.

The majority of those who die in fires, especially ones that occur in domestic dwellings, where most fatalities occur, lose their lives because of the effects on the body of inhaling smoke and toxic gases. Their skin might be covered in a greasy soot, but often such victims are not burned at all.

However, in those fires where there has been a large body of flame and great heat, and perhaps relatively low levels of heavy smoke, such as when accelerants are involved, victims can be burned to a point where they are barely recognisable. It is often hard to distinguish severely burned bodies from the general blackened debris all around. What appears to be a baulk of charred timber can, on close examination, turn out to be a body. In the worst cases, identification of the deceased is only possible by referring to personal dental records or, if possible, DNA samples.

***

As I have illustrated in the many incidents I have described, it was often the case that fatalities played a part in emergency call-outs. In the following short excerpts, I hope to convey further the breadth of the London Fire Brigade's responsibilities in all manner of situations and the gruesome sights that we came across on a regular basis.

Electricity at a Grave Price

We are about to sit down to an evening meal of sliced roast beef with horseradish sauce and a huge jacket potato, all washed down with scalding tea from the huge teapot, when the call bells toll. The Pump is ordered to a report of 'Smoke Issuing' on the adjacent station's patch, not too far from the London Hospital.

Within ten minutes, we pulled up behind the two local pumps in a seedy backstreet. It contains lock-up garages within the brick arches of a railway bridge over which a long freight train is rumbling past. There is a slight pall of smoke coming from a building opposite the arches and I sense a sickly sweet smell in the air. The huge bulk of the local Station Officer, his white combed helmet giving him another six inches of height, towers over several teenagers who sport the spiky hair of the fashionable punk movement of the time. The youngsters are strangely silent, their pallid complexions even paler than usual, and I deduce they are in shock.

'What you got, Ted?' Biff, our guvnor, enquires of his counterpart.

'It's a nasty one, Biff,' he replies quietly as he walks up.

I go into the front door of the building, a near derelict house that the punks appear to be squatting in, and turn into what was the kitchen, where I find three of the local station's crew gathered. The sweet sickly smell evident in the street is much stronger in here.

'The poor little sod,' mutters one of the three, stepping back from what is a large old electric oven next to the incoming service board for the electricity. Lying over the top of the cooker is a blackened and badly charred arm and the trunk of a teenager, the untouched head covered in the same spiky hair as the others standing mortified in the street outside. Two large strands of copper wire are under the youth and these lead back to the service board.

It will take an expert to work out exactly what has happened, but it seems evident that the youths had been illegally extracting electricity from a supply that had been shut down and it had gone tragically wrong.

We return in due course to our station and put into play the black humour that we often deploy as a way of coping with the stress of the horrific scene we have witnessed. Leading Fireman Dick Friedland, looking down at the plate of meat that is by now a wee bit overdone, says, 'This beef is nearly as badly burned as that poor little guy.'

'No bleeder would wish that sort of death on anybody, but they get what they deserve, messing about with electricity in that way, the silly bastards,' exclaims Barry Priestley, as his teeth sink into the overcooked meat.

An outsider might be shocked at such talk, but without its purging effect, which served to distance our sensibilities from the horror we had encountered, a lot of us might have ended up in a psychiatric ward, given the gruesome sights that our line of work so often threw up.

The Boy on the Train

It is just after 9 p.m. and we are at a rail maintenance depot alongside the surface electrified rail track. A local resident has called the brigade after seeing some smoke and burning in the depot's vicinity. I am with the guvnor as we walk towards the depot to investigate. The gates are locked and a large warning sign reads: 'DANGER – ELECTRIFIED TRAINS. DO NOT ENTER'.

Someone has placed a wooden plank, the sort used to provide a platform on top of scaffolding, against these gates at a steep angle. A radio message has already been sent requesting an official from the depot to attend, but if there is a fire in the compound we are permitted by law to enter.

We split the small alloy extension ladder into its two halves, placing one on the outside of the tall steel gates, the other inside them. In my hand I have a carbon dioxide extinguisher in case there is a small fire on anything still with an electrical current.

As soon as we get to the other side, an electric motive power unit carriage can be seen. On the roof next to the pantograph arm, which conducts the current from an overhead cable, a black pile of debris is gently burning, with the occasional flicker of a small red flame.

We take the ladder from the gate and I pitch it against the side of the carriage, taking care to go nowhere near any overhead wires. My first thought is that a short circuit has ignited some insulation on the pantograph. That thought is wiped from my mind by the shocking sight of the shape of a human leg. The debris is the badly burned and electrocuted body of a small person. About the only thing recognisable are the sides and soles of a well-known brand of trainer.

When severe burning is present, the heat can cause the skin to burst, while the tendons and ligaments in the upper limbs are foreshortened, producing the pugilistic pose of the boxer. Such an appearance is sometimes the first indication that a body has been found in a severe fire scenario.

Later that evening, we learn that the body we found was that of a 12-year-old boy who had been playing amongst the trains. Apparently, he had used the plank to access the depot and had managed to climb onto the roof.

There must have been a live current or the residue of one and in his playing he had made contact with it. He was electrocuted and burned instantly. I would not have wanted the task of the officer detailed to break the sad news to the boy's parents, surely one of the hardest duties of the police service.

Although death as a result of fire is not the way anyone's life should end, it was always easier to deal with the passing of someone who had reached an advanced age than to deal with the death of a baby or small child who had been denied the opportunity to make the fullest use of the life they had been gifted with. If a fireman with his own small children back at home becomes involved in the aftermath of a fire or other emergency in which small children have perished, it is very hard indeed to cope. It requires nerves of steel and an ability to pull down some mental shield to lessen the association with your own offspring. Not everyone has this ability.

Likewise, from my own experience, dealing with the most badly burned body is less difficult than dealing with someone who has died from the effects of inhaling smoke. In the former, the body is virtually unrecognisable as a human being, whilst in the latter humanity is all too obvious – the victim looks as if he or she is sleeping. Sometimes we would attend a fatal fire where the victim was hardly marked. As such it is easier to make a mental connection with your own loved ones and you have to steel the nerves and put such sentimental thoughts out of your mind.

A Deadly Candle

'*Smoke Issuing . . . Dock Street*' reads the ordering slip and we are rolling within 30 seconds. The incident address is so close to our station that we are there almost before we've completed pulling on our fire gear.

As we arrive, two Asian women clad in saris, about 30 or 40 years old, are waving frantically from the pavement outside a four-storey Victorian house. There is no fire or smoke visible in the street, but the BA men are poised, ready to rig should Biff, the guvnor, require it.

In fact, there is no need for BA or even a fire extinguisher. Lying on the floor of the rear room of a small flat is an Indian woman of about 70. She had been wearing a sari, but half of this is burned away, revealing her thin lower limbs. Her dignity is abused by her widely splayed spindly legs, probably a result of the way she had fallen. There are some brown scorch marks around her on a beige carpet and an extinguished candle in a small holder is on the ground near her. She is motionless, and there are no outward signs of burns to her body, but there is no pulse or breathing and her open eyes are as dead as those of a fish on a slab.

Biff requests an ambulance, as no matter how obvious death is to the eye of an experienced fire officer, we cannot confirm it; that is the physician's duty.

Biff takes some details for his reports from the two younger Asian women. It transpires that the candle was part of a prayer ceremony and somehow the flame had ignited the nylon-based sari.

A month later the coroner recorded a verdict of accidental death due to the effects on the lungs and heart of the sudden inhalation of fire fumes caused by the sari becoming alight. He mentioned that relatives should try to ensure that the clothing of those who use candles, especially the elderly frail, are fire resistant.

* * *

Whenever I was present at such incidents, especially if I had left for duty with some domestic tiff unsettled, I would be reminded of the futility of such fallouts. You are reminded by these scenes, where life is snuffed out, of how fragile our being alive is. I am sure that most of my crewmates felt the same and that at such times we all made an inner vow to never argue again. But most of us, similar to when we make a vow never to touch drink again when suffering a hangover, continue with our habits, as the mind shuts out the worst memories of your discomfort. We're therefore undeterred from a domestic dispute once the raw visions of what we have witnessed blur with time's passing.

As well as fires, it has been traditional for the UK Fire Service to respond to what we call 'special services'. These involve attendance at non-fire scenarios such as road, rail and aviation accidents, leakages and spillages of toxic or combustible materials, flooded buildings, persons buried by collapsed structures, persons threatening to jump from on high or even those trapped in lofty locations.

As fire and rescue crews are trained to wear BA, to work inside toxic atmospheres, at heights or in subterranean confined spaces, as well as being qualified to operate specialised cutting, lifting and spreading apparatus, it became inevitable that we would be first to be called to these non-fire emergencies and play a key role in rescue and retrieval operations. Since 2004, such non-fire attendances became a legal obligation on the fire and rescue service.

In the same way that victims of fire either appear unmarked or severely ravaged, non-fire fatalities can present a similar state in a lot of cases. However, it does not take too much imagination to work out that someone who has gone under a London tube train or is involved in a serious rail or air crash will usually have died from some quite gruesome multiple injuries.

Victims of high-speed road traffic collisions or the occupants of family cars where heavy goods vehicles are involved usually produce similar carnage. From my own experiences, it is often more difficult to deal with such trauma than those occasioned by fire and I believe this difficulty is related to the personal associations we form in respect of the place where a victim has lost his or her life. Almost every road traffic collision with fatalities that I responded to and that involved the family car and its passengers had an emotional impact greater than when dealing with victims of structural fires – it is the association with our own family car and the world around it that brings home that old adage: there but for the grace of God go I. The sight of familiar items, but torn, disfigured and bloodied within the wreckage of a car, strongly magnifies how, in an instant, in a second of inattention or error, life can end so quickly.

Chapter 9

More than Snooker, Darts and a Bed

It is our second night on duty and whilst the Pump's crew are out doing routine inspections and tests of some of the 500 hydrants on our ground, the rest of us are in the mess room on a Q&A session with Sub Officer Jack Hobbes.

Contrary to the popular beliefs of certain cynics, firemen do a bit more than playing darts and snooker and sleeping in between answering emergency calls.

The notion that fire crews do little but wait for a call to come in mainly stemmed from the so-called 'Phoney War', when the expected bombing Blitz on London from the Luftwaffe failed to materialise in the immediate period after war was declared. It had been the perception of some that those who had opted for fire rather than military service were taking an easier route and the terms 'scrounger' and 'parasite' began to be used.

No doubt there were some who considered that being a member of London's fire force might be a less hazardous route than the armed forces; however, once the bombing raids began, it soon became apparent that serving as a fireman, with nightly conflagrations to deal with amidst the deadly rain of bombs and flying shrapnel, collapsing buildings and burning gas, was just as hazardous as serving with the military. As a consequence, these same individuals soon became categorised as the bravest of the brave, and many memorials in London record the hundreds of individuals who made the ultimate sacrifice. Once hostilities ended, however, and the fire service reverted from the National Fire Service to local authority control, not all memories were sharp as to the inherent courage and sense of public service required of those who had chosen to serve in the highest-risk locations.

On top of this, the fire service is traditionally modest about its achievements; after all, it was known to many as the 'Silent Service'. It seemed to me that the mistake in this approach was in thinking actions spoke louder than words. The top brass of the time weren't convinced of the power of public relations, believing the courageous actions of fire crews spoke for themselves. The way I see it, it is a bit like a well-known prizefighter believing his exploits in the ring do his talking for him. If it is only by fighting him that you learn of his prowess, how do you let the wider public know? Likewise, if it were only through your experience with the fire service that you learned of the heroism implicit in its deeds, then as a wider concern it would lack public support.

The reporting of the work of the fire service 40 or 50 years ago was poor and this did little to counter the widely held beliefs within the public. What they perceived the service to be was very different from how the working firefighters saw it. Perhaps back then the top brass should have done more, although the annual review of the brigade at the Lambeth HQ, which royalty often attended, at least showed the high operational efficiency with hose, ladders and rescue lines. But none of this illustrated what went on in between responding to emergencies.

It had always been a tradition within the UK Fire Service that crews had to clean and maintain their stations and appliances, and this continued until the 1970s, when civilian cleaners were employed. Prior to that, we firemen had to do all of our own housekeeping. Sweeping and polishing floors, cleaning windows and scrubbing showers, sinks and toilets was the regular routine. But the trade unions had convinced management that it was demeaning for a professional emergency service to be employed as char ladies. They were probably correct, but those daily cleaning chores were once an integral part of our personal ownership of our fire stations, where so much of our lives was spent in between responding to the call of the bells. However, as fire and non-fire hazards increased in number, there was less time to involve ourselves with the chores of keeping the station clean.

A typical day or night duty included any of the following, once the meticulous start of shift checks of personal kit and appliance gear had been completed and recorded:

- Practical drills in the drill yard and tower, involving ladders, hose, pumping, rescue lines and lowering harnesses, road traffic accident extrication gear and breathing apparatus
- Standard tests on all items of equipment, which were segregated into daily, weekly, monthly and annual periods
- Cleaning of appliances and equipment. Hose washing, testing and stowage
- Breathing apparatus routine tests, checks, servicing and cylinder refilling
- Lectures and Q&A sessions on such diverse subjects as practical firemanship, hydraulics, water relaying, chimney and hearth fires, chemical spillage and containment, sewer rescues, sprinkler and automatic fire alarm systems, high-rise building fires, basement fires, underground railway procedures, first aid and resuscitation techniques

Brigade policy mandated that all crew received minimum periods of involvement and instruction in this wide range of topics, plus monthly attendance at the BA smoke and heat chambers and divisional exercises that simulated major emergencies.

We made periodic visits and inspections of selected premises for purposes of operational familiarisation or to check adherence to fire safety legislation. We also carried out pressure tests on the rising main systems required to be fitted to all buildings that exceeded four floors, checked scores of fire hydrants and toured the ground to learn the streets and the quickest route to any emergency call received.

So our time at the station between emergency calls was certainly more than snooker, darts and a bed.

Tonight Jack is testing our knowledge of the procedures involving lifts. Around the mess table are Bernie Rosenfeld, Toby Smollett, Martin Bauer and me.

'What's the difference between getting a call to "person shut in lift" and "person trapped in lift"?' asks Jack without at this point singling out anyone to answer.

New boy Martin has a stab. 'Isn't the first where somebody is just stuck between floors or at a landing but cannot open the door, and the second is where somebody is trapped and injured by the mechanism, Sub?'

'Spot on, old son. You'll go a long way,' says Jack.

'But don't go too far in a westerly direction, Martin, or you'll end up with those poseurs in A21 Hollywood,' jokes Toby with a grin, making a jocular reference to the 'A' Division's Paddington HQ and the other stations that contain most of the West End and its much visited tourist attractions.

'Take no notice, young fella,' says Jack. 'They're only jealous at your memory. Aren't you, Toby? Now, you can tell us all what the procedure is for "person shut in lift", can't you, mate?'

'I was only joshing, Sub. Anyway, the first thing is to locate where the lift car is and then—'

The bells toll, giving Toby plenty of time to compose his answer, as we pour down the 20-foot steel pole.

'*Fire. High Road. Pump called by radio*,' the duty man yells.

The road we are called to is no more than three minutes away and, as the Pump is inspecting hydrants at the opposite end of our ground and has been ordered on by radio telephone, we will be first on scene, but their two tones and bell can be heard in the distance.

No address has been provided, so we slowly cruise along the High Road with Jack giving the odd burst on the horns and a burst on the roof-mounted bell, whilst, with windows down, we scan and sniff for evidence of fire or smoke. The Pump charges up from the opposite direction and slows when it sees us; the two appliances stop in parallel.

'Just go up and down your side and take a shufty down the side streets, and we'll do the same this side, and if there is nothing I'll send a stop, Jack,' shouts Biff Sands.

Five minutes later we know it is yet another hoax call and are all returning, ready for the evening meal and a big pot of tea.

Once supper is finished, it is stand-down until 7 a.m. Tonight, Gus Witherington, a much-respected Station Officer who spent fifteen years here and who retired eight years ago, has called in to meet up with his old buddies Bernie Rosenfeld, Jack Hobbes and Dick Friedland, something he likes to do a couple of times a year.

A group of us are sitting around the long table in our first-floor mess room. There is a freshly brewed and huge pot of tea in the serving hatch to the kitchen.

'Great to see you again, Guv,' says Bernie, holding out his hand, still considering his former Station Officer as his boss.

'Hello, Bernie, old mate – but less of the "Guv" or young Biff will get me barred! He's your guvnor now,' responds Gus.

'Don't you worry about me, Gus,' Biff chimes in. 'The truth is you loved the job so much that you can't keep away, you old firedog. While I am here, you are always welcome, especially as you and your old mates here taught me most of what I know,' he goes on with a smile.

Gus Witherington had joined the brigade before the war; when he enrolled in the 1930s, many of the old timers had still been mariners. He could tell many a story about riding on the old open-bodied appliances – of how in the winter months they would often return from a working job with their wet tunics frozen so solid that they would stand up on their own. He would tell how their hands were almost stuck to the brass handrails and of how the short lifespan following their retirement was often hastened by the punishment their bodies had taken during such times. He had seen many of the conflagrations of the London Blitz and often enthralled us with his memories, once he could be persuaded to talk. He had lost a lot of good mates during the nightly raids in those dramatic and horrific times.

Gus Witherington had seen it all and nothing appeared to faze him. But what did concern him was the shift in people's attitudes, which he had first noticed in the so-called Swinging Sixties. His real pet hate, though, was the decline in discipline and the all-for-oneself attitude of increasing numbers who seemed to have little concern for the rights or feelings of others. Gus was a deep thinker with a very analytical mind, and a strong capacity for forward and lateral thinking, which left most of his peers at his rank standing still. I had heard some of the older hands say that he would have been an ace crime buster. He had noticed that common sense was going out of the window; governments were instead laying down laws, with little perception of how over the top things were becoming,

'You all want to enjoy the job as it still is,' Gus offered as he sipped the strong tea from a large white mug. 'I was reading somewhere about this new health-and-safety legislation and that the fire service will not be exempted from it,' he continued.

'What are you saying, Gus?' asks Jack.

'I'm saying that once this law starts to bite there will, in my view, be changes the likes of which many of us wouldn't believe. In my day, even yours, the maxim is to get into a job, even if there are no persons reported to be missing. It was only in the jobs that were so well advanced, such as a dockside warehouse alight from end to end, that we fought a fire from the street,' he rattled on with a passion driven by having been there in the thick of it countless times.

'I think that in due course officers in charge will become so scared of their own shadows, so fearful of messing up their promotion by getting a bloke killed or injured, that they will be too quick to pull you guys out into the street. Once you go down that road, you lose a lot of professionalism and with that a lot of your pride. I will even say that it might make the job more hazardous.'

'Why do you say that, Gus?' asks Bernie.

'Because Bernie, my old buddy, if blokes are pulled out before the seat of the fire is located, then a pound to a penny the fire will get away. That will lead to more chance of building collapse, a collapse that could be onto crews, and so you will have to forget the protecting of property, which has always been a natural part of fire service work.'

'You are probably right. Gus. But do you think that we should be risking our lives trying to save buildings when there are no persons trapped?' Toby Smollett, the watch trade union official asks.

'Look, old son, I hear what you're saying and, believe me, I know what the union's stance is on crew safety. I was once doing your job myself,' Gus responds. 'But the tradition, from recruit training onwards, was always that you were taught your job is to save life, protect property and carry out other humanitarian duties. In any case, I have been to enough fires when those outside have been adamant that everyone has been evacuated and a normal, precautionary trawl has found people and sometimes they were already beyond help. On top of that, what are you going to do at, say, a deep-seated basement job where it ain't practical to pump in high expansion foam? Leave it to burn? We would be there for weeks, old mate. So if we start to fight fires from outside, we could see a lot of damage to premises and their contents that could have been reduced by getting in and locating the fire's source.

'You've got to remember that a burnout puts people out of work and this affects the local economy. The insurance companies will put their premiums up and we'll all end up with dearer goods. It's common dog, ain't it?

'I'd even lay a bet that in the future firefighting from outside will see a reduction in the number of stations and pumps, and of course that will mean fewer firemen. But listen to me going on. You must all think I'm a bleeding dinosaur,' Gus concluded, quickly draining his mug and then wiping his eyes, which had become slightly moist.

The old firedog Gus was so strong in his convictions as a result of his massive practical experience. This would forever convince him that such a huge metropolis needed to be in possession of a force not lacking in anything that might hinder its operational effectiveness.

In an ideal world, organisations and even governments would see policy decisions on public protection from fire being made only by those in possession of a comprehensive practical experience gained at the sharp end. But as in so many other walks of life, things were changing on a regular basis and even practical experience seemed to be becoming taboo in the eyes of some. Some of these shifts were beneficial, but the concerns expressed by people like Gus, that the noble ethos of the fire service as life and property protectors was being slowly eroded, seemed to run in parallel with the feeling that the value placed on a human life was being lost.

To men like him who had fought a war and given so much to protect humanity, any suggestion that the often dangerous work of fire and rescue could be lumped in with other non-emergency local authority services, and people's lives subjected to ill-thought-out budgetary costing, was unthinkable. In fact, it was almost treason.

Chapter 10

Gus Witherington's War

During lulls in the action, which even the busiest stations experience, often the diversion of a spaghetti western or the pages of a good book sufficiently kill time – but not when old sweat Gus Witherington is visiting. No, there is little on television or on the printed page that can keep us enthralled like the tall, twinkling blue-eyed old firedog, as he regales us with stories of his life in the brigade before the war and his experiences of the Blitz and in the post-war years.

Born in Hoxton, then known as one of the country's most deprived areas – a far cry from the trendy artist studios and galleries it housed in later years – he soon moved with his family to Poplar in the heart of the docklands, that area at the head of the U-shaped loop of the Thames familiar to those who watch *EastEnders*.

In the early 1930s, when Gus joined the brigade in his early 20s, it came under the auspices of the London County Council (LCC) and consisted of a Northern Division and a Southern Division, which were subdivided into geographical districts north and south of the Thames. Ten years or so before he enrolled, horses had only just been retired from their tasks of hauling some of the brigade's escape ladders.

Not long after he had become a recruit, one of the largest dockland fires of peacetime occurred at Butler's Wharf, near Tooley Street. The raging fire had swept through a seven-storey warehouse crammed with a variety of highly combustible goods. Some 50 engines and 100s of men were needed to deal with it, during what he told us had been a freezing cold day.

Previously a cabinet maker for a small company in Shoreditch, near Old Street, where the headquarters of C Division stood, he was encouraged to join the brigade by an uncle who was at the time serving with the fire service. Following basic training, he was posted to Station 73 Euston, one of the 13 stations across the capital to house a 100-foot extended Turntable Ladder. After three years there, he was posted to Station 2 Manchester Square, just off Park Lane, and a couple of years later was posted to Station 28 Whitechapel, where his long association with the East End and its fires began. That move coincided with the outbreak of war.

Gus was always ready to point out the strictness of the brass, and the rawness of the conditions of those early years, something that we found hard to envisage. Gus used to tell us that firemen in the early 1930s worked about 72 hours a week.

The first enclosed body appliances were not introduced in numbers until later in the decade and he remembered some of the older men referring to the open-bodied Pumps and Pump Escapes as the 'pneumonia wagon' on account of the health risks of returning sitting unprotected in the cold in fire tunics that were water-soaked following firefighting operations. He also remembered those old sweats talking about the times when men didn't sleep in a dormitory but on trestle beds in the appliance room, almost fully clothed – for them, it was considered a slow turnout if the front wheels crossed the door's threshold in more than 30 seconds.

We found it hard to believe that up until 1936 firemen still wore brass helmets. The number of electrocutions had led to the design of a cork helmet capable of withstanding up to 11,000 volts of electricity, which could remain intact at up to 350°F.

Gus would speak movingly about the way many residents of the East End had to live during the pre-war years; indeed, well into the 1960s. He recalled many of the tenements were infected with bugs and were overcrowded and lacking in sanitary facilities relative to the size of the buildings. He had attended many fires in such properties, as well as those in dockside warehouses crammed with all manner of merchandise, from alcoholic spirits, rubber and grain to wine, clothing, rags and timber, to name only a few.

From 1939 until 1940, the non-arrival of hostilities led to the 'Phoney War', but when, on the night of 7 September 1940, following aerial dogfights over Kent, Surrey and Sussex, the Blitz raids commenced, there was, said Gus, nothing 'phoney' about what he and his colleagues witnessed. 'The Blitz was incredible. It continued for about eight weeks and what I saw wound me up so much that I still get angry about it today, three decades on.'

He recalled one raid, begun at five in the afternoon, when over 600 aircraft dropped both high-explosive and incendiary bombs across all of east London. 'Jerry was aiming for the vulnerable targets, like Beckton Gas Works and the power station at West Ham, as well as the Woolwich arsenal, Millwall Docks, Limehouse Basin and the massive dock warehouses at Rotherhithe,' Gus explained, with a look in his eyes that indicated what he had seen was still very clear, such had been its powerful effect.

'The Germans were crafty so-and-sos,' he went on. 'On one occasion, they waited until the tide was out before starting, knowing that we would be unable easily to pump from the river.'

He said that although the government tried to hush things up, so as not to demoralise people, the truth was that on the first night of the London Blitz 436 men, women and children were killed and more than 1,500 seriously injured.

'You would have had to be there to see the conflagrations caused by the air raids. An awful lot of the procedures we use at major fires were based on the lessons learned during the Blitz. I remember at one inferno the street was running with hot molasses from a nearby warehouse. The heat radiated was unbelievable. That's where we learnt the value of setting up a water curtain to absorb the radiated waves of heat.'

Before the Blitz, the Auxiliary Fire Service (AFS) was ridiculed. Firemen, Gus told us, had been called such names as 'skivers', 'card sharps' and 'parasites'. But once the raids started, perceptions of the AFS completely changed: 'They knew there was as much chance, if not more, of copping your lot in the East End as there was in the armed forces,' Gus explained.

'I'll never forget the night I was working holding a jet in Limehouse with a mate. We lost water and he walked over towards the trailer pump to find what was the cause just as a high explosive came down. I must have been completely in the opposite direction to the blast and, although deaf for a few days, was fine. Old Matey was never found – atomised, I should imagine, poor sod – and married with five little nippers,' he said, wiping a rough hand over an eye. He had also experienced the V1 and V2 flying bomb/rocket assaults after the Blitz that caused much carnage and property destruction.

Within a few years of the wartime NFS (National Fire Service) being returned back to local authority control in 1948, he made Station Officer, seeing service throughout the East End up to the late 1960s, when he was forced to retire.

There were quite a lot of old firedogs like Gus around when I first came to London and one of them, Cyril Demarne OBE, wrote a fine record of what life was like during those terrible nights and days of the Blitz in his book *The London Blitz – A Fireman's Tale*.

In 1991, Queen Elizabeth the Queen Mother unveiled the Firefighters Memorial in Old Change Court, to the side of St Paul's Cathedral, which commemorates all those in the fire service who died due to enemy action, some of whom were Gus's best pals.

The Albert Embankment, the Lambeth headquarters of the brigade, opened a few years after Gus became a recruit, also had a memorial in its reception. This commemorated all who lost their lives on duty during peacetime within the capital, and some of the names on that list were his mates also.

Perhaps it was the effect of those losses that ensured Gus had been a much respected guvnor and 'good hand'. He certainly enlivened many of our night duties during those periods when the call bells were silent. He will always be remembered, like so many of those strong characters of those exciting years.

Chapter 11

With New York's Bravest

Not very long after my transfer, I picked up a book in a north London library entitled *Report from Engine Co. 82*. Its author was Dennis Smith, son of Irish immigrants who had made the often rough Atlantic crossing to start a new life in New York City back in the 1950s. At the time of its publication, Smith had spent several years as a fireman, one of New York's Bravest, as they are so affectionately known. The fire service is revered in the city, a result of acts of tremendous courage and life sacrifices made during its long and proud history guarding the population from fire and calamity.

Every year since its inception in 1865, the Fire Department New York (FDNY) has seen on average eight firemen lose their lives on duty.

Dennis Smith's book was his account of his life as a fireman in the crime-ridden area of the South Bronx, where he had served for about six years. The book became a million-plus seller, being translated into several languages, and told of the high drama and the social conditions of a run-down part of New York in which over 9,000 responses to fires and other emergencies were being made every year. No doubt the busiest fire and rescue unit on the planet.

I was unable to put the book down and my reading of it coincided with a TV documentary called *The Bronx Is Burning*, which was made by a film crew who rode out to the many fires that were occurring, often incidents of arson where landlords wanted to remove tenants or because of issues connected to the drug wars going on then in the Big Apple.

Such was my enthusiasm for New York and the courageous exploits of the FDNY, and so keen was I to try and learn from their experiences, that I managed to secure a low-cost charter flight during their freezing winter, no doubt a reason for the low fare. Even today I remember this trip like it was last week, although my introduction to the FDNY came about in a very different way from the one I imagined.

I was in East Harlem, a largely Puerto Rican quarter back then, and a down-at-heel part of Manhattan located north of Central Park, which itself was then a notorious spot for muggings and assaults. As I walked around, it was hard to keep upright without slipping on the compacted snow and how I wished I'd brought my climbing crampons with me!

As everyone knows, the New York days and nights are filled with the wail of police, ambulance and fire sirens, so when about five of the huge green 'Neighbourhood Police Team' liveried cars sped past me, weaving on the icy roads, this, I thought, was not unusual. But as they screeched to a rapid stop and a cop holding a pistol in hand leapt out, my immediate reaction was that an episode of *Kojak* was surely being filmed. I was wrong.

At that moment, I passed the red single doors of a tall firehouse (some FDNY firehouses accommodate just a Pumper). I saw the faces of a handful of firemen pressed against the door's small square glazing panels, looking intently out at the police action. A small wicket door was open and, not knowing the sort of reception I would receive, I went in and announced who I was and where I worked.

The fire and rescue service is an international family in so many ways and when I mentioned my connection to the London Fire Brigade, it was like I had provided the combination to a lock behind which lay a friendly world of very direct but good-humoured banter. Within seconds, a mug of steaming coffee was being poured from a percolator that seemed to be permanently on the go in this dingy firehouse whose walls were painted that same dirty mid-green seen in the *Kojak* episodes of Manhattan South Precinct.

To cut short a long story, it turned out that the police action was a result of the shooting of an off-duty New York cop. I was told he had been having an affair with a Puerto Rican woman, apparently, and had told her that he was ending their relationship. On hearing this, she had taken his revolver and shot him in the spine! So this was true-life drama – not the filming of a TV series, as I had imagined.

Since I was alone (Carmel was back home at work), I had only myself to worry about and such was the hospitality of these men that they offered me a bed in the upstairs dormitory, something that would never happen back in the UK, where, even then, there were strict protocols on non-members of brigades staying on local council premises. I called the Edison Hotel in Times Square and let them know I wouldn't be back that night.

Not only was I given a bed but I was also allowed to ride out to calls, all unofficial, the one safety proviso being that I rode inside the Pumper rather than 'riding the back step', which two or three of the crew did back then. (The system was revised when a number of fatalities and serious injuries resulted after men had been thrown off.)

We responded to about eight calls that night: some, like everywhere in the world, false alarms; others, to small fires in the huge local project housing blocks.

I well remember the difference between the American and British experiences – the 'in your face' New Yorker stance compared to the much more starched-shirt approach I was used to – as illustrated that evening in the firehouse, in between calls, with the old steam radiators banging as they warded off the sub-zero temperatures outside.

Earlier in the day they had dealt with a small blaze in a bagel store. Their reward? A couple of baker's trays of bagels and doughnuts. I must have been seen eyeing these tasty morsels, as the lieutenant in charge of the night shift, a huge man with the build and facial resemblance of American actor Brian Dennehy, suddenly said, 'Listen, Buddy. I see you looking at the bagels –

well, if you want some, help yourself, pal. This ain't stuffy England here.' But in a friendly way, with a big grin!

The next morning I mentioned that it was my plan to visit the home of Engine Company 82, the subject of Dennis Smith's book, and Ladder Company 31 in the South Bronx, the fire station affectionately known as 'the Big House', or 'La Casa Grande', so termed because it held three or four appliances and, like Harlem, had a large Spanish-speaking community. With the words 'You wanna be careful, buddy, walking around up there; it's a bad, bad area' ringing in my ears, I thanked them for their advice and their huge hospitality and took the subway north.

Given that the Big House was responding to almost 200 calls per week – that's about 30 every 24 hours – and of them many were protracted firefighting and rescue jobs, it was not surprising, therefore, that when I finally arrived via the icy pavements from the subway through this menacing district, with many buildings scarred by the signs of serious fires, all the three appliances were out; the familiar *Mary Celeste* scene, of kicked-off shoes, empty coat pegs and boot shelves, greeted me instead. Fortunately, within minutes Ladder 31 returned – I hadn't forgotten the friendly warning from the fireman back in Harlem, or Dennis Smith's descriptions of the South Bronx as one of the worst areas of the whole of the USA for vice, extortion, drug addiction, shootings and street muggings – and I could breathe a little easier as I introduced myself to the lieutenant in charge of the huge aerial ladder that had featured so prominently in Smith's fine accounts.

The welcome, once they learned of my London connections, was no less than that in East Harlem. It felt almost surreal, seeing that only a few days earlier, during my flight, I had re-read *Report from Engine Co. 82*, to be actually inside the station on Intervale Avenue in which the book is based.

A few minutes later the actual Engine Company 82 returned, with a crew of blackened-faced men who had put out a medium blaze on the top of a seven-storey tenement. I remembered Smith's description of the different physiques of the men and of how the big 'elephants' – the 6-foot-plus, 250 lb guys – were usually the ladder men. Theirs was one of the most hazardous of roles in a job, and in a city, that was hazardous by its nature. These big guys had the task of not only rescuing from upper floors but also ventilating the smoke and heat. The construction of many of the buildings was such that if the heat wasn't vented, a fire could enter the roof space (or cockloft, as they termed it) and prove very difficult to suppress, with the risk of severe lateral spread.

As a result of the high incidence of drug-related crime, residents would often fit tough locks and bars to the doors on the roofs of tenements to prevent break-ins. The downside of this in a severe fire is that delays are created for the ladder men, who have to force a high-level entry to effect a rescue.

In the UK, these 'above the fire' ventilation techniques – which included using hand tools and powered saws to cut vent openings in the roof directly above the raging inferno – were not used, even in some of the bigger central London buildings.

Many of the eight or so firemen who had paid the supreme sacrifice each year while working for the FDNY had been ladder company men who had been plunged into the hell below when a fire-weakened roof had collapsed under them. Every time a man died, the bell signal 5-5-5 would sound across all stations followed by an announcement of the name and number of the man or men who had perished, and the fire alarm street box number where the fire had occurred. It was the awareness of this very real potential for death in the job that led many stations to mount on the wall a placard with the sobering words: 'Tonight could be the night'.

The smaller men usually made up the Engine Company crews. Their task was also search and rescue, but primarily it was their task to 'stretch' hose lines into the building, and to locate and extinguish fire.

So busy was Engine 82 that at the time of my visit a new system was being tried of bringing in relief Pumps and crews for the busiest parts of a duty to afford some respite for the men at the Big House.

Hardly had E82 got back and had a wash, then replenished the onboard water tank and serviced their BA sets, than the bells tolled for yet another call. I didn't need a second invite to grab myself the visored helmet, rubberised dog clip-fastened fire tunic and thigh boots off the shelf of off-duty crew and clamber into the rear cab of the engine. I really did think this was a dream. Here I was, on the actual appliance in the title of the book, hearing that rise and fall of the wailing siren and the tremendous braying of the air horns of Ladder 31 behind as we weaved through the mean streets of South Bronx. So busy was it then, and so often did the local populace see tenements blazing, that they didn't even turn their heads as we roared past.

The call was to back up crews on a neighbouring patch and as we sped along I saw plumes of brown smoke rising from another serious tenement fire that had nothing at all to do with the incident we were racing towards, such was the business in those years – a period the FDNY would later call 'the War Years'. As it turned out, we were not needed and returned to Intervale Avenue.

Again, as at Harlem, I was given the chance to stay on overnight but because it had been snowing heavily, and most of the crew travelled as much as 60 miles each way into work, some of the day tour couldn't go home and some of the night tour could not get in, so there were no beds. No problem, they said. If I didn't mind sleeping across the expansive cab seats of Ladder 31, I was welcome to stay.

As the snow thickened, the Ladder had snow chains fitted to its huge wheels. I don't know what made me think that I would get some sleep as I pulled an FDNY blanket over me at about midnight. This was the busiest firehouse in the world.

Within 20 minutes, the box numbers started to toll. Crews knew if a gong beat related to a box or street alarm on their patch, and on each duty a man spent a few hours in a small office monitoring and confirming signals and filling in a log book. At about one in the morning, I was stirred from a fitful doze by the duty man's shout: 'Box 7879 . . . Atlantic Boulevard' or similar. Then I heard him yell: 'And the chief goes too.' The latter meaning that the Battalion chief, a bit like the ADO in London, and who was provided with a huge red car, would be required to attend.

Within seconds, the half-open cab of the ladder behind me was taken up with two monsters of men, their eyes bleary from a sudden awakening.

Even with the snow chains on, the huge vehicle slewed across the snow-covered highway, just missing the vertical iron support columns of the elevated railway track. For a brief second, I thought of the awful consequences for Carmel and myself should we collide with something. Not insured, unofficially riding on a FDNY appliance . . . but the drama soon made me forget, as it can when you are still in your 20s and feel immortal.

Within minutes, we were halted at the street alarm that had been pulled, but there was no sign of fire or other emergency. At the time, a lot of these mechanical 'pull alarms' were being replaced with a voice communicator, however as a large proportion of the South Bronx is occupied by Spanish-speaking residents with poor English, language difficulties could cause confusion on a call-out. In addition, if there was any sort of emergency – fire, drug overdose, street or domestic altercation, suicide, childbirth – a box would be pulled. This was because in such a teeming city, with police and paramedic services so stretched, the one outfit that you could guarantee would turn up was the FDNY.

The Ladder and Pump toured the streets slowly, the compacted snow eerily muffling the sound. Nothing. The lieutenant rewound the clockwork mechanism and we set off back. Within seconds, the radio squawked 'Ladder 31 – the box you have just attended has been actuated again.' On with the red beacons, wailing sirens and we return.

A tiny man with the look of a Hispanic, wearing a raincoat that almost touched the snow, was walking from the box, seemingly grinning. About two years earlier one of the colleagues of the Big House had died after being thrown from the back step of a Pumper responding to a hoax alarm from a street box in this location. The grinning proved too much for the temper of one of the huge men in the back, a man who would not have looked out of place as a forward player with the New York Giants. His open hand was that of two men in size, so even the very gentle slap he administered must have stung, if not split the man's lip. I caught the end of a string of invective that threatened murder if he pulled a box again for no reason.

Maybe the tiny man had a reason but did not have the language or hand gestures to display an emergency. Whatever, I felt that, unfair or not, there was one man who might think twice before sending men on a wild goose chase in an area where the all-too-frequent fires of the time meant men responding to a hoax call might otherwise have been saving lives. It was, of course, the memory of a good fireman lost that had forced this piece of rough justice.

Before I bade farewell that next morning, we went out to another eight hoax calls at various locations before picking up a serious blaze in a motor tyre fitter's below tenements. I was astounded to see residents refusing to evacuate the upper floors. I was told that such was their poverty, so few were their worldly goods and so high was the incidence of burglary and robbery that many would rather risk death by fire or smoke inhalation than evacuate and possibly lose their TV or other possessions to thieves.

My absorbing visit to spend some time with New York's Bravest would forever be etched on my mind. Of the many things it taught me about my work, one of the most important was to always expect the unexpected when serious fire is about.

Perhaps my most significant memory, however, related to the deep love of life that I found to exist within the majority of those who chose to stand a vigil by night and day in that teeming city. It seemed to me that it was that profound sense of humanity, born out of that love of the life we are given, that was the major motivator behind the heroic actions that have so often seen individuals make the supreme sacrifice whilst trying to preserve the lives of others.

Each and every year the FDNY holds its annual Medal Day. At this historic event, after deep deliberation following the study of front-line reports of heroism, gallantry and consummate public service, the chosen recipients for some 61 medals are presented with their awards of recognition in front of family, friends and workmates. One would have to be a person of extraordinary insensitivity not to be stirred by the written citations that are printed within the handsome official programme that accompanies the ceremony.

The following is an abridged citation detailing one such act of gallantry, which took place during the early hours of a winter's night in one of New York's high life-risk cosmopolitan districts:

> Shortly after the units arrived at the seven-storey residential block, he saw all five windows on the fourth floor spewing flames which were curling up into the fifth-floor windows. As he was about to enter the tenement, piercing screams from an evacuated resident indicated persons were trapped. As he made his way up the staircase, he could smell the unmistakeable odour of a serious fire and sensed that fear which is often at its strongest just before entering, but, thanks to his training and unit pride, he suppressed that natural apprehension. The door to the affected floor had been left open by residents who had made a valiant but fruitless rescue attempt. This allowed heavy brown smoke to enter the landing, making it very dark and extremely hot.
>
> Flames were leaping from the doorway and rolling across the ceiling and this created waves of great heat that began to burn his neck and shoulders as he crawled towards the rear rooms of the apartment. At the end of the corridor, he observed a huge enveloping body of fire and, being by now twenty feet into the corridor, realised that there was no alternative escape route for anyone in this building. Notwithstanding this, he persisted in his crawl in this superheated atmosphere with visibility almost zero into the bedrooms in his search for the man reported missing. Then he caught sight of an unconscious person half in and half out of a bedroom doorway. He grabbed the man's arms, half crouching to achieve this and to better be in a position to drag the man out, and the raising up of his body put him into the even hotter smoke.
>
> The heat was so fierce that he could feel the skin from the man's arms peeling off. Undeterred he continued to drag the victim back the twenty feet to the landing, where he was assisted by other crew members. In carrying out this gallant act, he exposed himself to a heat which caused third degree burns to almost a half of the victim's body. He unhesitatingly

> placed himself in a most hazardous situation with no alternative exit route and sustained a number of serious burns, but still continued, so that the 48-year-old stranger might live, and we are proud today to award him with this medal as recognition of his outstanding bravery in the line of duty.

When the news broke on 11 September 2001 of the atrocities at the World Trade Center, my mind went back instantly to my thrilling days in the city that never sleeps. The FDNY lost 343 firefighters in this attack; this meant that in a few short and horrendous hours the FDNY had lost as many personnel as in the previous 40 years. I have little doubt that some of the tough men with tender hearts, certainly the younger of those who had greeted me with such friendliness 30 years earlier, were amongst those who failed to return from the blackest day in the department's fine and noble history. For them, the sober message on the wall placard sadly came to a horrible fruition.

In October 1966, a few months before my own career began, the biggest disaster to involve the FDNY until 9/11 took place at East 22 and East 23 Streets in Lower Manhattan. Twelve men perished, consumed by an inferno raging unseen under what was thought to be a solid concrete floor of a drugstore; in actual fact, it was concrete laid on a timber floor.

Throughout my long career, I never forgot the lessons of that awful tragedy. Its teachings about how crucial advance information is ahead of an emergency to the safety of the public and the fire and rescue crews were undoubtedly essential to my development within the profession.

In 2010, I was able to pay my respects to fallen comrades at Ground Zero and at the memorial for East 22nd Street. I am ever conscious of my good fortune in surviving unscathed after 30 years on the front line, whilst those fine individuals who keep that eternal vigil perished.

Chapter 12

A Close Shave

'That street is as flammable as a load of petrol in a celluloid container,' shouts out Leading Fireman Dick Friedland as he clambers up into the rear of the Pump at just after four in the morning on a brass monkey of a February night.

He'd caught sight of the location of the fire on the teleprinter roll as he dashed past night duty watch-room attendant Ricky Tewin, who was waiting for the machine to stop its metallic chattering before tearing off two copies, one for Biff Sands, the other for Jack Hobbes on the Pump Escape.

The East End street to which we have been called at this godforsaken hour is infamous for the amount of serious fires that have occurred over the years. Like so many parts of the district, it is a powder keg comprising garment manufacturers, reproduction furniture factories, timber yards, cheap and seedy bed and breakfast establishments, to name but a few. Such is the potential for conflagration if fires are not quickly located and suppressed in this area that the automatic ordering to all calls to fire or smoke issuing, or even to a smell of burning, consists of three Pumps, plus a 100-foot Turntable Ladder.

Our station is smack bang in the middle of this atmospheric quarter, streets away from where Oswald Mosley's fascist blackshirts once strutted. The honours and awards board on the wall of the watch room bears witness to the high courage displayed down the years by its personnel, who have braved the killing flames and searing, choking smoke to save members of their community.

'I'm sure I can smell the bleeder,' exclaims the six-foot-three Lofty Morphard from the rear of the Pump Escape, his long legs – the reason for his nickname – crammed tight up to the BA set brackets positioned on the partition between the rear and front cab.

'I think you are right, Lofty,' shouts Jack Hobbes above the diesel's clatter, as ex-collier Barry Priestley clasps the wheel with his thick fingers, steering the eight-ton engine around a junction. 'Be ready to rig in BA, fellas,' Jack bellows. 'The smell is getting stronger by the effing second.'

I grip tightly on to the bracket of my own BA set, as I glance across the rear cab at Lofty and Rick who, along with Jack, Barry and myself make up the five-man crew of the Pump Escape. The crisp light of the fast-rotating beacons of the Pump in front illuminates the cab in pulsations of its electric blue light. The so-familiar and peculiar odour of a burning building is now much stronger and the uncertainty of what awaits us at an hour where a fire might have been developing and increasing its fiery hazard for a long time tenses our features; that same strobing blue reveals apprehensive expressions under our helmeted heads.

No matter how many occasions I have responded to emergency calls – and by now it must run into a few thousand – the vague, stomach tightening knot of fear that this might be the job from which we don't come back, the 'Tonight could be the night' of the FDNY, is ever-present when, as now, clues exist that we are responding to a fast-developing blaze. I have read enough military books, and dabbled enough in human psychology, to appreciate that this feeling is the fight or flight element within our minds and that, at times like these, it would be a bare-faced lie to state that you'd rather be here than in a safe, secure situation elsewhere.

My readings of the two world wars, and the wars in Korea and Vietnam, have indicated clearly to me that I am not alone in feeling such a powerful urge to not be there. Some military psychologists and psychiatrists have put forward the argument that some of the least worried soldiers are those with a tendency to be psychopathic in nature, men who feel little fear or remorse. I met few, if any, firemen who, at least outwardly, showed any such tendencies. The 'flight' effect is a natural one when you are aware that death can come from a number of causes. All that's needed is a sudden flashover of superheated flame, or a wall or floor to collapse, like it had on New York's East 22nd and 23rd streets. The unpredictability of that moment left me wondering if my ambitions were grounded more in the young man's sense that he will live forever than in the cold, hard facts of life and potential death of such a hazardous occupation.

Yes, it would be easier to have a less vivid imagination, but the rigour of training, the pride in the badge of the brigade and not wanting to let mates down better ensures you learn to steel yourself and control your morbid thoughts. That way we learn to face whatever fire the gods throw at us.

Those gods must be in a bad mood tonight because as we career round the corner into the infamous street, Jack snarls out, 'Effin hell! We've got a right bastard of a job here!'

Angry and massive forked tongues of darting red and yellow flames are shooting out of the top-floor windows of a six-floor building on the front wall of which a large placard announces 'Colour Printers and Lithographers'. For a brief instant, my mind thinks back to Alf Cooke's and 'that fine body of men', which seems light years ago, even if it is not. The lurid light from the flames – just like that of the Calls back in Leeds – makes the night seem like day, as we shudder to a halt, well clear of the structure, in case a wall weakened by the searing heat is about to topple.

However brief my career in printing, it was long enough to know that if you had to choose a building that would burn well, then you couldn't do much better than a large printing shop. There were reams of paper, bound in brown wrapping, the sort of stuff that produces choking smoke; huge vats of a paraffin-based cleaning liquid we called blanket wash, used to dilute printing ink from rollers and cylinders; muslin, cotton and cheesecloth, some soaked in dried ink and solvents; plus a host of chemicals used in the production of printing plates. Altogether, a very unhealthy recipe, should fire strike.

Add to this the huge weight of presses, guillotines and pallets of unprinted and printed stock, some above the ground floor, and it is difficult not to want to flee, especially with an imagination as vivid as mine.

'Get your BA men started up,' comes the stentorian shout of Biff, shaking me from my morbid thoughts. I take the 30 lb-plus oxygen set from its bracket, glad that I had not skimped my checks of its cylinder content and gas tightness at the 6 p.m. roll call some ten hours earlier.

I heave the set's harness over my head and pull the body belt tight, my eyes on Ricky Tewin. 'It's going like a bastard, mate, ain't it?' he remarks, his brown eyes wide open, his foxy features illuminated by the intense shimmer of ruddy light from the inferno into which we will have to venture.

'You're not joshing. And we need to be really on our guard here, mate. I used to be in the print and there's as much hazard in here as anywhere.'

'Thanks for your reassuring words,' Ricky mouths, before we insert mouthpieces and clip into the head harness, double-checking the cylinder pressures as we hand our tallies to Lofty, who has been designated what we call 'Stage One BA Control'.

The plastic tallies – details of our rank, name and pressure – enable him, using a table on his control board, to track when our warning whistles will sound and the time we are due back into the fresh air. If we don't show, it is his responsibility, after informing the officer in charge, to send in an emergency team to locate us.

I never go into a hairy fire like this without thinking of the London blokes who have lost their lives in BA jobs, and from which our BA safety protocols have been derived.

Following the deaths of firemen at Covent Garden fruit and flower market in 1947 and Smithfield Meat Market in 1958, improved BA control procedures were introduced. These procedures gradually developed and only a few years earlier a new guideline had been issued, advocating use of knotted tags to help crews find their way out. The trouble with this was unless you could find a secure attachment for it every few yards, it soon became a 'right bunch of bastards'. As a result, we tended not to use this approach, relying instead on the time-tested practice of keeping a hand on any hose line taken in and by feeling for the couplings, using them to tell you if you were going in or out.

Yes, learning of fellow firemen who have made the ultimate sacrifice in order to do their duty did bring about the 'flight or fight' instinct. But I have yet to be at a dodgy job where a man has refused to do his duty. Of course, we wouldn't get a bullet in the head if we refused – but the UK Fire Service did have its own disciplinary regulations at the time.

Such is the serious size and developing speed of fire-spread that Biff has sent a '*Make Pumps Eight*' and while donning our sets, another station's crew has laid out a charged hose line and used a sledge hammer and large axe to force the locks of the heavy double entrance doors.

The radio squawks and its tinny loudspeaker echoes a message:

'*From Station Officer Sands at Dock Street – printing works of six floors, about 120 feet x 80 feet. Whole of top floor alight, remainder of premises smoke-logged, BA and one large jet in use.*'

Ricky and I are low on our haunches as the doors are forced, in case the inrush of air creates a deadly fireball that would consume us. No explosion – yet! However, a cloud of dense, rolling brown smoke, tinged with yellow streaks of sulphur, a nasty sign, can be seen. The smoke is so thick you could almost cut it with scissors. True to form, and without the protection of BA, smoke-eater Biff is crouched at our sides.

'I want you to work the jet into the ground to find the fire's seat,' he says. 'I got an idea that it might have started on the ground floor and spread up to the top floors via an open-sided lift. I've been here on a visit in the past.' He's as cool as the proverbial cucumber, quite befitting a man who had seen mates who were complacent go toes up in Korea 20 years back – a salutary lesson to him on having a critical approach.

With bosses like Biff, you didn't need the threat posed by the national discipline code to do your job; he's a real 'guv' and we'll follow him into the centre of any situation if he has risk-assessed it as being possible to penetrate – it is his deep and broad practical experience that enables him to know when it is time to press on and when it is time to beat a hasty retreat.

As I enter the building, the heat reminds me of that felt when your bare limbs are too close to the honeycombs of a gas fire. But in this case it isn't just on my legs, it is everywhere, enveloping me like a red hot blanket. The blue melton cloth of the fire tunic and the rubberised leggings over the thin blue overall-type trousers below it give scant protection or insulation. Within seconds, perspiration is pouring from every pore.

I am holding the hand-controlled nozzle, crawling to get under the heat band, whilst tugging the heavy charged hose that Ricky humps behind me. Outside other men are slowly feeding in what is our lifeline. It is a lifeline to protect us from the flames that we attack, but a lifeline to the exterior should things take a turn for the worse. As I crawl, my mind turns once more to Alf Cooke's and the combustibles within that printing factory. I hope and pray that we can soon locate the seat of the blaze and get this powerful jet onto it, to kill the fire dead before it does the same to us.

I'm wearing a small pair of pink-coloured rubber-framed goggles to give protection from the blinding hot smoke, but even after smearing the inside of the lens with a proprietary 'nil-mist' they soon steam up on the outside because of the humidity. Every so often, I wipe a tunic arm over the exterior lens and poke a finger to clear the inside. I have just done a clear-up, which lets me see shimmering, dancing flames over to our right. I instinctively open the nozzle and aim the jet at the ceiling above, from where the water cascades onto whatever is burning and the steam briefly darkens the fire's light.

Biff's hunch has proven to be correct. The flames have got into a vertical shaft and roared up it like a rat on fire going up a drainpipe. This is what has ignited the top floor and it is highly likely that the intermediate floors could soon be involved also, not a pleasant thought when you are deep inside such a hazardous environment. At such times with no persons, such as a night watchman, reported missing, it is easy to start questioning if it is worth risking life and limb to save property alone. There are two answers. First, one can never be 100 per cent sure that there are no persons inside. There have been more than a few occasions across the country in which a failure to carry out a search has resulted in a body being found later. Second, part of our unwritten creed includes the saving of property that is someone's business or an employee's livelihood.

The representative body would prefer we didn't unnecessarily risk firefighters' lives in saving property where no persons are trapped, as its role is tied in heavily to the safety of its members, but, in relative terms, we would not have a very strong argument for an enhanced salary if all we got paid for was life protection. On many stations, including some in the quieter outer suburbs, there are few fatal fires and therefore to argue the case to retain an expensive fire and rescue service on the one-in-ten-year prospect of its need is a difficult one to sustain in financially stringent times.

I don't want to convey the impression that such thoughts are to the mind's forefront when struggling to suppress a rapidly spreading fire, but around the mess table the topic was being increasingly raised, especially as the 1974 health-and-safety regulations became imminent.

I'm moving the jet around in a constant rotating and dipping arc, knocking down the flames and cooling as much as possible of what is being devoured by their ravenous red tongues, when suddenly I feel Ricky's hand tugging at my right arm.

I shut off the powerful stream to reduce the crashing noise and turn to face him, gingerly lifting the edge of my goggles, as it is still exceedingly hot, to rub the steamy vapour from the lens in an attempt to see him in the ruddy light.

As we flounder in the heavy smoke, I recall Mr Bexon, my school's French master, reminding me that the word inferno derives from the French word for hell. How apt, I think, as I strain to make out what Ricky is saying. He is grunting something around the edge of his mouthpiece. This is a dangerous practice and is called 'talking aside'. Our BA instructors warn of its menace with a 're-breather' set, which re-circulates used oxygen mixed with fresh from the cylinder on our backs. 'Talking aside' involves slightly moving the mouthpiece to one side to permit clearer speech. This can allow the deadly carbon monoxide to get into the circulation and fell you without warning. If Ricky is taking this risk, something very serious is amiss.

'Look up at the friggin ceiling' is the half-clear message, and I see his right hand pointing upwards. My already racing heart misses a beat. Smoke and flames are now travelling very rapidly across the ceiling away from the lift shaft and going up through what appears as a hole. This is, of course, the underside of the weighted floor above our heads.

It takes me a split second to pull my own mouthpiece to one side. 'Turn around, grab the hose line and follow it out! Now!' I bellow as best as I can.

We have penetrated some 50 feet into the ground floor and in the gradual advancement, aiming the jet, slowly feeding the heavy hose, it has taken some 15 minutes. With the very urgent need to get away from the gaping hole above, we cover that distance in about three minutes flat.

It isn't a second too soon. An almighty thumping crash shakes the building as a portion of floor, and whatever it was supporting, falls onto the very spot from which we have just so speedily hot-footed. Clearly, the great heat of the fire on the old floorboards, impregnated with years of old solvent, grease and ink, has weakened them to a point where they have failed. The weight of whatever has collapsed caused a massive surge of superheated smoke and gases that almost blasted us out into the street.

Guvnor Biff had been checking things at the factory's rear when, as he informed us later, he heard the terrific crump and knew instinctively that some sort of structural collapse had occurred. In spite of his bulk, he sprinted back to the front, showing his concern for his men.

'You fellas OK?' he asks breathlessly, as we report back to BA Control, where the divisional BA Control van has taken over from Lofty, and close down our sets. Once satisfied that we are unhurt, he arranges a head count, which confirms that no other men were inside when the floor gave way.

It is a close shave, and we are lucky to have escaped, but part of our mission is to save property. This incident is par for the course of being a fireman in the early 1970s.

Fate must have had a hand in our escape. You can be in possession of all the practical experience in the world, and be led by the most trusted and competent guvnors, but fire is a devious, unpredictable beast when out of control and can snuff out life like breath on a candle flame.

Chapter 13

The 'Good Hand'

As mentioned at the outset, after working in the north of England I joined the London Fire Brigade to satisfy a long-held ambition to work within the capital city.

I have no doubt that it was the high level of anticipation of the uncertain – the not knowing what the next emergency call might be – that exists on the busiest stations that played such a large part in my motivation to serve on them.

The high number of calls and working jobs also meant that the many hours of drills and simulated scenarios that I'd undertake could be done in the sure knowledge that within a tour of duty the potential for putting them into play on the real-life stage was far more likely than on the far quieter stations in the outer suburbs.

It was almost as if the words printed on the cover of a national Sunday newspaper of the time, 'All human life is here', applied to the environs of the most active inner-city stations.

I eventually found myself serving at Islington Fire Station, a redbrick Victorian structure on Upper Street only a few hundred yards from the teeming Holloway Road – the southern end of the A1 Great North Road, down whose undulating route I had first travelled to London. Its many calls to street after street of houses, tenements and factories, its underground stations of Essex Road, Angel, Highbury and Islington, and the neighbouring high fire-risk areas of Holloway, Stoke Newington and Shoreditch, is providing me with some first-class operational experience – the kind the London Fire Brigade considers very important for career-ambitious personnel.

'Ops experience' is the watchword, both in terms of the 'watch' of fire stations and in the minds of senior ranked officers who sit on promotion interview boards.

Most higher ranked officers have very solid operational pedigrees, but times are changing; some of the younger high-fliers seem to be trying to change all that and appear to put style over substance . . . though none of them are there to meet me when I report on a cold January morning for a promotion board interview. My interrogators are old school and will possibly grill me under a very bright lamp!

In a fire brigade that had experienced such catastrophic conflagrations as those caused by the Luftwaffe's bombing blitzes and which had in peacetime to deal successfully with some of the nation's largest blazes, it is natural that practical competence on the fire ground should be the order of the day (and, more often, the night!).

Theoretical knowledge is essential to pass the written parts of promotion exams, but woe betide any would-be junior officer who has not gained himself a solid practical pedigree before arriving to command a crew or a watch on any of the 'sharp end' stations.

On these busiest inner London fire stations, some of which are responding to over 3,000 calls a year, many being solid 'working fires', the efficiency and personal safety of firemen relies upon an officer in charge being a 'good hand'. To be known as a good hand carries great weight and gives the fireman credibility based on his practical ability to keep cool under pressure and to exercise sound judgement in the firefighting tactics employed.

The 'bookman' – the 'paper officer' who panics when people are screaming for help four floors above the street, when attending a heavy fire in the early hours, say – will be ridiculed and condemned as incompetent. In short, on the sharp end stations, he is the sort of man who will be eaten up for breakfast and spat out!

To avoid such personal embarrassment and the risk of attracting a derogatory label that can stick with you throughout your entire career, a prudent fireman with aspirations to command does all that he can to earn and win his spurs. He builds a solid reputation on the foundations of coolness, courage, operational competence and the ability to handle men.

My interview is at Brigade Headquarters on the Albert Embankment at Lambeth. It is an impressive building of nine storeys, with a seven-bayed appliance room, all overlooking Old Father Thames. To my impressionable young eyes, this late 1937 building, with its commanding position in full view of the Houses of Parliament, appears to be everything that a fire brigade headquarters in one of the world's most famous cities should be.

I am shown to a side room, one of many on the second floor, where large dark polished wooden doors and wall panels are in abundance. There is a good 15 minutes before my interview, so I try to relax by looking at an impressive display of photographs of some of London's most notable fires over its long and proud history. There are images from the 1930s of brass-helmeted men fighting a docklands warehouse fire in temperatures so low that huge icicles are hanging from warehouse walls and brigade ladders. From the 1960s, there are dramatic shots of a rescue by hook ladder from a tall building within the Square Mile from where grey rolls of smoke, looking like roughly wound balls of knitting wool, are issuing.

There are images of major fires in the affluent West End and, by contrast, a grim photograph of a fire-bombed minicab office in the East End's Hackney, with barely recognisable blackened bodies – taken during a terrible period when the famous black cab 'Hackney' carriage trade was feeling threatened by the inception of the private hire vehicle.

Looking at these photos it occurs to me, even several years after first inhaling that pungent odour of stale fire smoke that seems to permanently permeate the fire tunics hanging in the gear room, that the buzz I had when I first stepped into the inner-city firefighting environment, which I thought to be the initial thrill of a newcomer, has a permanence that time or a thousand emergency responses will never still.

'Come!' a booming voice answers, in response to my knock on

the large polished door, which bears a shiny brass plate naming one of the London Fire Brigade's upper-echelon officers. I march over to the far end of a cavernous office. Behind a large, green-leather topped desk sit the two who will grill me for the next 15 or 20 minutes. Their backs are facing the Thames, on which I can see tugs and lighters bobbing past.

'Sit down, please,' commands the senior of the pair in a rasping voice that sounds as if his vocal cords have been cured in the thick smoke of too many fires. Behind his gold-rimmed reading glasses, the dark eyes are set in cavities around which are weathered brown pouches, the result of many years of smarting eyes in those cruel seas of smoke in house, flat, factory and warehouse. I can almost imagine piercing those brown bags under the eyes and smoke remnants oozing out from a thousand inner London fires.

Both of these 'top brass' cast their critical gaze over my bulled-up black shoes and the razor sharp creases of my uniform, but as important as a smart appearance is to a disciplined service, they are far more interested in what I know. Operational experience is being sought in promotion candidates: it is the recommendation reports that are key and evidence that a man seeking promotion is up to the highly responsible task of commanding crews safely and of protecting the capital's buildings and its populace.

But I must have fielded the difficult googlies of the complicated fire-ground scenario questions that he and his Divisional Commander colleague bowled because a month later my name is placed on the year's panel from which promotions to Leading Fireman will be made.

But for now, it is back to being a fireman at the sharp end.

* * *

As usual I arrive at the station a good 30 minutes before the six o'clock roll call. I use the first quarter of an hour to clean my fire gear, polishing to a sheen the black calf-length leather fire boots, the black rubberised leggings, the black webbing belt, axe pouch and black helmet that resembles those worn by Roman centurions.

Although the more recent design is much more compact, with

a slightly serrated comb and a rear projection to better protect the head, the helmets worn by the longer serving men, including our own guvnor, more closely resemble those of Victorian times.

Over many years, through countless black-and-white press photographs, the London Fire Brigade's image has become synonymous with the traditional fire gear of black helmet, black double-breasted tunic, boots and leggings. Those black items of protective apparel contrasted with the scrubbed-clean white ash wood shaft of the axe and matching white coiled line attached to the belt, in addition the white helmet denoting the ranks of Station Officer and above.

Photographic archives display these piano keyboard contrasts of uniform at numerous fires and emergency scenes across the capital city. They evoke a thousand memories and stand as a strong reminder of the Brigade's history and its presence, not only for those who have spent time within its ranks but also for those who have either lived and worked in London or visited its bustling streets and quiet, secretive backwaters.

None of these images are more poignant than those illustrating the major fires in which firemen were killed: such notable incidents as Covent Garden (1947) and Smithfield Meat Market (1958), as already mentioned, but also the fire at Dudgeon's Wharf on the Isle of Dogs in Millwall (1969) – five men were killed when hot cutting operations on disused tanks of flammable materials went horribly wrong – and the 1974 rescue carried out at a Maida Vale hotel-cum-hostel, where a probationer fireman was killed along with six of the employees. Several others were seriously injured.

The black-helmeted heads contrast with the strained white faces of men, which reflect the sadness of their loss in streets shrouded in funereal wreaths of life-taking smoke.

With my own fire gear now prepared for the roll call, I carry it up to the appliance room. At just before half past five, permission is given for me to relieve one of the day watch, so I go into the watch room to ensure that the duty man enters the change of rider in the station log book.

It is in the watch room that the teleprinter is located, where the card indexes showing the routes to all streets on and around are filed, and where, on its wall, a large map spreads out, showing the whole of the station's ground.

Above the teleprinter, a large brown clock shows that there are some fifteen minutes left of the nine-hour day duty. Most watch rooms face the street and during the last half-hour of a duty it is customary, even though Brigade orders forbid it, for men to congregate and amuse themselves by watching the passing scene – and to be the first in line to collar an oncoming man so as to get off early.

'Bloody hell, mate, you're keen, ain't ya? There ain't a night goes by without you relieving some fucker. What's up, mate? Has her indoors kicked you out?' asks 'Motormouth' Ricky Tong with a sneer, nonchalantly rocking back on the tubular steel chair on which he is sitting adjacent to the large street-facing window.

'Something like that, Ricky, but you'll have noticed that I never relieve you,' I respond with a half grin, knowing all too well that to reveal that one of the main reasons for my coming on duty early – to try to get a larger slice of the action – will be a passport to constant ribbing.

However, it is the case that such accusations only tend to demonstrate the irony of the situation. If the truth be known, probably a good three-quarters of every watch, certainly on the busy inner-city stations, are equally as passionate about their work. But we like to assume an air of indifference and wear the 'old green waistcoat' to disguise the fact to our crewmates, certainly during station stand-down hours. It was out on the fire ground, though, that this enthusiasm was evident.

During my time in the East End, there was one man who, to all intents and purposes, was Mr Cynicism personified and I have never forgotten him. His whole attitude and demeanour when not involved in an emergency response was negative. He grumbled about the Brigade's Principal Officers and he was always moaning about appliances and equipment. He basically tried to convey a lack of interest in his work. However, come the urgent clamour of the call bells, he underwent a Dr Jekyll transformation. There he was, first to the appliance, fully rigged in his gear before most of us had our boots on. And at fires he was unbeatable in getting his hands on the nozzle.

During a post-incident debrief and amidst the informal 'yarning' around the late-night mess-room table, the following story was revealed concerning Mr Cynicism. I remembered the fire, occurring one Saturday lunchtime in one of those mean streets near a point where the boundaries of Shoreditch and Whitechapel overlap, involving one of those tall Victorian terraced houses so typical of inner London, but the finer details of his role in lifesaving were only later divulged.

As the Pump Escape and Pump roared urgently into the wide street, a mother was screaming hysterically for her young son to be saved. The mother was surrounded by a small crowd, their necks craning up to the top-floor window from which thick clouds of that ugly yellowy-brown smoke were billowing.

Getting rigged in the BA set was a slow business, as I have said. When individuals were involved, many firemen just slung the set over their shoulders, jammed in the mouthpiece and got into the building to begin a search, but in this case Mr Cynicism didn't even look at a BA set. He ran straight in. Straight up the stairs. Ducked under the dense smoke and fast rising heat as he made his way to the top floor, where the boy was reported to be. A long corridor ran off the top landing and he dropped flat, his nose just above a grubby, threadbare trammelled carpet, sniffing out that thin layer of breathable air. He saw a glow – a red torch of flame made a dim orange by the masking filter of the choking, rough-edged smoke that would soon overcome him if he didn't find the boy and get out fast.

He pulled himself with his arms and elbows down the long corridor, like an SAS soldier crawling under the nose of the enemy's barrels, his head spinning as he reached the bedroom door. He crawled across the floor until his hand found the leg of an iron-framed bed. He felt on top. No one there. Felt under the left-hand side. No one there. Then he heard the soft whimper of a sobbing child. He felt further around under the bed, totally blinded by the brown smoke that now had the heat of a furnace within its layers. On the right side, his large, rough hand brushed the soft smooth skin of a young child.

In a second, he had the boy in his grasp, pulling him by the wrist across the floor. His strength was almost gone now, sapped by the debilitating, life-taking hot veil of fire fumes. He knew he wouldn't make it back down the stairs. His mind raced as the black shutter of coming unconsciousness began to fall.

He was in the centre of the room when, into his fast-receding consciousness, he heard the pure cockney accent of his old firedog instructor at Southwark 15 years earlier: 'If yer lost in smoke, stop. Find a wall. Work around it to find a door or window.' A life-saving tip of good firemanship that he now struggled to implement.

He headed across the floor, pulling the now unconscious small form. His hand hit a skirting board, brushed an electric power point, then the long drape of curtains around a window. He felt up and reached the wooden inner sill, knelt up, still grasping the boy's wrist; he daren't let it go, might not locate him in the horrible, hot, lethal fog.

Then on his knees. His left hand un-studded his personal axe pouch and grasped the steel blade. Fierce swings at the glazing with the flat of the blade brought shattering glass, then fast-rushing smoke out into the street. Further blows with the axe allowed the smoke to temporarily thin and the sweet light of day and life appeared.

He pulled the limp form of the lad up and held his head out into the life-giving London air. Within seconds, the escape ladder was crashing into the wall below the window and strong hands took the boy to the ground, to a waiting ambulance and a waiting mother; he hung out of the same window, pulling in lungfuls of oxygen, vaguely aware that the worst was now over, before his colleagues brought him down to safety.

Never again, I thought, after hearing of that heroism would I make assumptions based on limited information. That man had been the kind who when the chips aren't down delights in running against the grain, but who, when they are, like on that Saturday, come into their own, defying all predictions as to their actions. He was indeed a 'good hand' and was deservedly awarded a commendation for his sterling efforts carried out like so many in those days without the protection of breathing apparatus so as to save precious seconds – often the difference between this world and the next. Which it was for that boy, who thankfully survived because of Mr Cynicism's selfless actions.

Back to the watch room, and there is no extra 'action' for me on this occasion, in those 20 minutes or so before I parade with my watch. Following the routine handover checks of appliances and their inventories, and of breathing apparatus, we sit down to a mug of hot sweet tea, a Brigade-wide tradition, before starting on the evening routine, which tonight is the continuation of the technical lecture, interrupted by a call the previous day.

The subdued ring of the bell that sounds whenever the teleprinter starts to clatter out its message mingles with the babble of conversation around the mess-room table. Although the 'printer' conveys a number of routine items of information, such as the lists of out duties for the next night shift to cover station crews depleted by sickness, its alerting bell is also the usual prelude to the clamour of the large red call-out bell on the mess-room wall. As a consequence, the subdued ringing causes us to instinctively tense, cocking ears and placing senses on alert should an emergency call be coming in.

After about 20 seconds with no bell, we normally assume that the message is of an administrative nature.'It'll be the out duties for—' Johnny Carres begins, his voice drowned out by the urgent clamour of the call-out bell.

'Bleeding hell, that was a delay,' shouts out Mike Deveen, as he and the rest of us spring up, pushing back our tubular steel chairs with a collective scrape and rumble on the polished wooden floor, briskly but calmly heading for the pole house and the 20-foot drop to the appliance room below.

Chapter 14

'Jumpers'

She cannot be more than 18 years old, and her ginger-red punk style spiked hair is already sodden with the persistent spray of drizzle falling over the capital on this early spring evening. The Pump has been called to '*Person threatening to jump*', an emergency known to the London fireman as a 'jumper'. The venue is one well known to both ourselves and the residents of this district as a spot from which more than a few people have chosen to end their lives.

She is standing on the outside railings of a cast-iron bridge above a busy major road. Along this route, the headlamps of the late evening traffic reflect on the shiny black surface of the wet road – traffic that includes the so-familiar red London Routemaster buses, as well as the ubiquitous black cab, plying their trade in this thick sea of vehicles. The young woman had been seen climbing over the railing by an elderly resident of a nearby flat. She then called the police, who in turn called the Brigade.

A short, squat inspector is waiting at the roadside, his black raincoat glistening with rain under the yellow light of the tall lamp standard directly above him. In an accent as broad as the River Clyde, he says, 'It's a young lassie. She's threatening to jump and she'll need more than an aspirin if she does. It must be 70 feet down to the road.' He speaks in a dour way, with the deadpan expression of a case-hardened copper who has seen more human drama in his twenty years with the Met than many see in two lifetimes.

It certainly is a long way down.

If she jumps, there is the obvious threat to others below and it is becoming clear that the shut-down of the carriageway directly below would be a wise decision.

'I've tried to talk to her, but she told me to eff off,' says the inspector. 'She's asked to speak to a female PC, so I've requested one,' he goes on, and as he speaks a small white van pulls up. A tall, youngish-looking woman gets out and the inspector trundles his heavy frame along to speak to her out of earshot of the young girl.

'Geoff,' says Station Officer Biff Sands, 'request a Turntable Ladder and send the informative: *One female threatening to jump from road bridge – police in attendance.*' Geoff Joynt repeats it all and radios it to Control.

The Turntable Ladder can be extended up to where the girl is perched and, hopefully, she can be persuaded, notwithstanding her agitated state of mind, to get onto it and make the assisted descent down the 80 feet or so of hard metal rounds (rungs). Alternatively, she can be helped back onto the roadway. The inspector returns and informs our Station Officer that he has instructed that the incoming carriageway be closed, and that the WPC will try to reason with the girl.

'Do you think it's wise to risk your colleague's life, going over onto such an exposed position, especially with how greasy it will be on those cast-iron supports?' inquires our Station Officer.

'But she said she wants to speak to a woman,' the inspector replies.

'Yeah, and I understand that. But I think one of my men, who is used to working at heights, ought to get down to her first and try to reason with her. You never know, but she might be OK talking to a fireman.'

'Very well then, perhaps we'll try your way. You never know,' replies the inspector with an air of resignation, tinged with a grudging respect for the work the Brigade carries out.

'Right, I want a volunteer to put on a safety line. I want to get out to try to reason with her and persuade her that she has to allow us to assist her,' our guvnor says quietly.

'I'll do it, guv,' calls out Mike Deveen. 'You know the ladies can't resist me.'

'Typical effing Mike! He'll do anything to pull a bird,' quips Niall Pointer.

'The only pulling I want is her back up here on terra firma,' grunts the guvnor, with a grin that belies his concerns for both the girl and Mike.

The rescue line is 130 feet long. At its end are two adjustable legs into which a person can be secured. The ultimate plan, if she won't return the way she has gone, will be to extend the Turntable Ladder and persuade her to get into it. She will then be assisted down to the ground. But if Mike can secure her in the harness, if she tries to throw herself from the stanchion, she will be safely held.

Working swiftly, nimble fingers secure the free end of the line to a suitable stanchion whilst Mike places one leg of the harness over his head and places it under his armpits. Pulling it tight, thinking all the time of the standard tests carried out only a few days ago on the rescue lines back at the station and of how the line had passed its test with no signs of weakness.

If ever proof were needed of the validity of such tests, tonight is a prime example. If he manages to persuade the girl to place the sling around her, then she jumps, she might pull them both off – then the line will really be tested, especially as his plan to have one sling on the girl and one around himself is unconventional, both slings normally being on the person being rescued.

Before climbing over the bridge railing, Mike takes off his helmet. He knows that some people in a very disturbed state can over-react to anyone in uniform. It's almost as if the authoritarianism suggested by tunics, caps, buttons and belts comes over to them like a red rag to a bull. So anything that makes the situation more informal can be of assistance in winning an individual over.

Mike Deveen, with the rescue line being held taut by his crewmates above him, gingerly makes his way across to the huge girder on which the girl has climbed and is precariously perched about 30 feet from him. Far below, he can see that the incoming traffic has been stopped and only the northbound lane is open.

About 100 yards along the incoming highway, Mike can see the red of the Turntable Ladder, discreetly parked, awaiting the instructions by radio from the officer in charge.

'Go away! Leave me alone! I don't want to talk to anyone,' the girl shouts out in the thin, nasal, monotone accent of West Yorkshire, as she suddenly notices his presence.

She is about 5 ft 3 in. tall, very thin and pale, and clad only in a pair of tight-fitting denim jeans and a short-sleeved T-shirt, all of which are thoroughly sodden by the incessant drizzle.

The powerful street lamps above are bright enough to reveal the tell-tale scars and scabs on her right arm of countless punctures by needles, where she has searched for a vein to feed her 'mainlining' addiction. That scrawny arm is now encircling the cold, green-painted cast-iron stanchion. Apart from a similar girder on which she is standing, shivering with shock and cold, this is the only thing that is keeping her from a sudden frenzied leap onto the unforgiving road so far below.

'Eff off!' she screams. 'One more move and I'll jump – what have I to live for now?' she sobs, her words carrying on a breeze.

'My name's Mike, what's yours? You can tell me, just take your time. I can help you get down from here, and someone will be able to help sort things for you. It's too wet and cold up here, and whatever you're feeling like now won't be half as bad once you're down and enjoying a nice hot cup of tea. So come on, love, what's your name?'

She remains silent, her teeth chattering, and Mike wonders what depths of despair would lead someone to place herself in such a situation. Thwarted in love? Deep in debt? Grief, anger, resentment? Maybe the malevolent effect of mood-altering drugs, or the irrational thought processes caused by a mental illness?

Mike moves very slowly closer to her, pausing often to try to get her to speak, to divert her mind from the suicidal thoughts she is experiencing. 'Tell me what it is that's bothering you. Remember, a problem shared can be a problem halved. Just tell me at your own pace and I promise you I will listen, but please tell me your name.'

After what must seem like an age to Mike, she turns very slowly to face him. Her eyes are puffed up with crying and her teeth are chattering.

Then, in a tremulous voice, she blurts out how she had come down to London after a row at home with her parents at their home in east Leeds. How her dad was a coalminer until a serious accident had laid him off work and led to his enforced retirement. Soon afterwards her mother had developed breast cancer and had had to leave her job. They had berated her for failing to hold down a steady job and after a couple of her mates had gone down to the capital she had traced them and they had allowed her to share their bedsit until she had got some work. She had got a job serving in a pub in King's Cross, where she had met a guy whom she had moved in with. He introduced her to drugs. Now she was an addict. Only this week, her already collapsed world had plumbed to even deeper depths. Not only had her boyfriend told her to leave, but also, during a phone call home, she had learned from her sobbing dad that her mother had died a week earlier.

'What's the point of living? I've nothing to live for,' she sobs, looking down at the wet road 70 or so feet below.

'You might not believe me, but there's plenty to live for. And what about your dad? I reckon he must need you now as much as at any time. What state will he be in right now? And if you were to come to any harm, he'd be devastated,' says Mike quietly. 'Look, all we want to do is help you, it's our job, and there are loads of other people who can help also. Anyway, I'm cold enough with my thick tunic on, you must be frozen solid. Just let me get up to you and I can help,' Mike continued, all the while feeling the secure tautness of the rescue line paid out from above. 'Come on, let me make you secure with this line, then we can get you down to the ground and find some warm and dry gear and a nice hot drink, yeah?'

The girl doesn't say anything for several minutes. Mike knows from past experiences that people in her state of despair can in one second lead you to believe they have listened and will comply, only to leap to their death in the next second without warning.

She moves her scrawny, scarred and goose-pimpled arm a little higher up the wet iron stanchion. She then half turns her right foot and peers far down to the carriageway, as if contemplating her own fate were she to leap off her footing.

She turns her white face, with its now sodden spiked red hair, towards him and whimpers, 'How can you get me down?'

'No problem, but first let me get up alongside you, yeah?'

'OK, but hurry. I feel sick and faint.'

Mike tugs on the line and, looking up, can see the Station Officer's white-helmeted head peering over the bridge rails. 'Give us a bit of slack, guv – I'm moving up alongside.'

He feels the line slightly slacken and slowly edges his way up to the girl, the rest of the crew above following his every move, ready to take up the slack and hold him should he slip.

Then he is alongside her. 'OK, you are going to be fine. Just do everything I say.'

Mike Deveen's thick fingers gently grasp her right upper arm. He knows that if she tries to jump now, then at the least with her puny frame, a quick tightening of his grip should hold her. But he has to get the second webbing sling on to ensure an even more secure situation.

'Take this sling in your left hand and, while I'm gripping you, put it over your head and push your left arm through. We also want it around your waist and back. Then I will keep hold of you as you put your right arm through. Then we'll pull it tight.'

'Yeah,' she replies, 'but I can't go back on t'road,' she sobs, a look of terror in her eyes as if pursued by some demon.

Slowly, nervously and with trembling hands and legs, she does as she is instructed. Mike breathes an inward sigh of relief as he tightens the harness, feeling an almost simultaneous tug as the crew above take up the slack line. They are now secure. If she jumps, she will have to overcome not only Mike Deveen's vice-like arm-lock around the stanchion, but the massive breaking strain of the rescue line.

'Ladders, guv, she won't come back over,' he turns and voices to the guvnor.

From far below, a diesel engine fires up and the huge set of 'Ladders', which from this height looks like a Dinky Toy, moves forward from its stand-off position at the roadside. The huge silver-grey sections of the ladder slowly elevate and then extend, the metallic click of the pawls echoing eerily as if counting down the seconds. Within a half minute, the top extension, with a Leading Hand secured on his flip-down foot platform at the head, is slowly reaching up towards Mike and the young girl. Almost an hour has passed since the Pump had first arrived and Mike himself is feeling ready to get off this exposed position high up above one of the capital's major arteries on such a wet and cold evening.

'Rest,' the Leading Hand says into his microphone, instructing the operator 70 feet below that the ladder is at the correct height. 'Move in towards the bridge about 18 inches.'

'Eighteen inches it is,' squawks the confirmatory message from below.

Very slowly the ladder comes in, and with a superb piece of precision operating, aided by the Leading Hand's instructions, rests no more than an inch from the stanchion on which the rescued and rescuer stand, and at an angle to facilitate the hazardous shift from the bridge stanchion onto the top extension, where a strong pair of hands awaits. Both Mike and the Leading Fireman know that nothing is certain in this sort of incident until the ground is safely reached. Accordingly, the safety line, although slipped from Mike, is kept on the girl and paid out under tension during the slow, aided descent, while he is assisted back onto the roadway by his mates' strong hands.

Mike's arm had been around the girl's shoulders, shielding her as best he could from the drizzle and cold breeze, before helping her onto the ladder. But however chilled she is, it isn't the irreversible icy coldness of death that could have, and unfortunately had in the past, been the outcome witnessed too many times at this location.

Within seconds, the wet, black roadway becomes closer. Then they are down on the welcoming tarmac. Met by an ambulance crew and the policewoman, who herself had so gallantly offered to go out on that exposed position to reason with the irrationality of the troubled mind. But for now, at least, the young exile from east Leeds is safe.

* * *

About a year later we are on night duty and have just returned from yet another false alarm. It is early summer and still light. I jump down from the Pump and notice a petite young woman standing nervously on the station forecourt, a few feet from the appliance bay doors.

'Can I help you? You seem lost,' I say, walking across to the doorway. There is something vaguely familiar about the woman, but I can't put my finger on it.

'I wonder if you can help me find the fireman who helped me once?' she says quietly in a distinctly northern accent.

'What did he help you with, and do you know his name?' I ask.

'I don't know what he is called, but he helped me with a bad time I was having about a year back.'

'Well, you don't have to tell me if it's a personal thing, but it might help me help you if you could explain things a bit.'

'Oh, it wasn't that. I got very down at that time,' she goes on, looking down at the ground, as if embarrassed to continue. 'It was on the high bridge up the road. I was really out of my head with a lot of things. I wasn't bothered what happened and would have jumped off if it hadn't been for that fireman bloke. He was so nice and understanding, and I really do want to say thanks to him,' she blurts out, seemingly glad to at least have told me why she has been waiting here.

It was not every day that we got called out to persons threatening to jump off high bridges and I remembered it at once, especially the part played by Mike Deveen on that evening. As it turned out, Mike had transferred out of the Brigade a few months back. He had met a girl on a walking holiday in the Lake District. She was from Blackpool and he had transferred to a northern brigade to be closer to her. In any event, the personal details of Brigade staff were never given out to anyone, although it was clear that for anyone to have known about the fine details of that incident of a year ago they were likely to have been genuine.

I recalled how around the mess table after that job Mike had filled us in with all the details, including that the young woman was originally from Leeds in West Yorkshire.

'Where are you from?' I ask of the pale, thin woman who looks as if a few solid meals of Yorkshire puddings with potatoes and roast beef wouldn't do her any harm.

'Up north, Leeds . . . Look, do you know this fireman, can you help me find him?'

'Yes, I do know him, but he has moved on.'

'Do you know where he's gone. Have you got his address?' she enquires. 'It will mean a lot to me to properly thank him, as at the time I was so spaced out, what with the drugs and other things.'

'Look, the best thing you can do, if you really want to thank him, is to write to our personnel section at our headquarters at Lambeth. If you do that, they should forward it to him, as I think they should know where he is.'

'Don't you know where he is? Can't you let me know?'

'I know that he moved to another fire brigade outside of London, but I don't know his new address. He was a great bloke, but we weren't off-duty buddies or anything. Just hang on a minute and I'll write down the address to write to at Brigade Headquarters.'

'Oh, OK, if that's the only way, then that's what I'll have to do. Thanks for helping.'

I hand over the Albert Embankment address. 'I hope that's OK. Sorry there's nothing else I can do to help. And, by the way, how are you getting on today?' I ask.

'Things are getting better a bit, but I've got some way to go. I live in a hostel at the moment and London can still be very lonely at times,' she replies.

'It sure can, but I was on that call that night and I reckon that the fireman and others who helped you knew what they were doing. It will be good for you to locate him. So do write to that address and best of luck for the future. I'm sure that things will all come good in the end. Take care of yourself now.'

She folds the scrap of paper I had given her and places it in the top pocket of the denim blouson jacket she is wearing. 'Well, I better be off then, and thanks again for your help tonight. And also at my bad time last year.'

'That's all right. It's what we're here for. Mind how you go.'

And she turns and walks away into the fading light.

A few years after this brief encounter, I was at a retirement party for one of the first Station Officers that myself and Mike Deveen had served under when at Islington. Like many of the retirements of those days, the respect in which most of the 'good hand' guvnors were held resulted in a large number of colleagues past and present turning up to say their own farewells. Given that Mike had moved a few hundred miles north, we were all pleasantly surprised to learn that he had found the time to make the trip.

It was good to see him again and we spent a most enjoyable evening exchanging yarns about the many working jobs we'd attended during those busy days and nights at Islington. Our conversation inevitably included that night, when, high above a north London road, he had used his natural charm and real professionalism to convince that distraught and unbalanced young girl that she had far more to live for than she was able to envisage at that point.

It turned out Mike had eventually received a letter from the young woman, and several others over the intervening years. How satisfying to hear that she had gradually got away from her addiction to drugs and had found new happiness with a guy she had met back in her native Leeds, where she eventually returned to be with her dad.

In the last letter, Mike said, she was now happily married with two young sons. She had said that her own new life and contentment would never have come about but for his efforts and those of the others present on that March night.

The fire and rescue worker's job is such that it is not often that we are able to discover how life turns out for those involved in the wide variety of emergencies and human dramas to which we respond.

How satisfying, then, to learn of how that shivering young girl, who on that night had been inches away from death as a result of the distorting prism of addictive drugs, had regained her rationality and rediscovered the will to live – and that the heroic efforts of Mike Deveen had not been in vain.

The author during the time he was serving with the Wakefield City Fire Brigade, 1968.

In the uniform of the West Riding County Fire Service, 1970.

At the London Fire Brigade Training Centre on completion of the transfer course, 1972.

With New York's Bravest, alongside Engine Company 35 East Harlem, 1973.

INSET: An everyday type of fire in the South Bronx, 1973. Note the tenant still on the fourth floor.

The aftermath of the same fire, which took place in a tyre depot, with tenements above.

Hook ladder training, 1970s, showing a fireman 50 feet up, leaning back as he enters the window. No safety nets here! (© LFB photos.com)

INSET: Hook ladder training – a two-man, two-ladders drill. Note the others far below. It's a long way down. (© LFB photos.com)

Getting into a smoky job from the head of the trusty escape ladder, early 1970s. (© LFB photos.com)

An example of the type of fire that occurred during Gus Witherington's day: a rescue using the escape ladder at a serious blaze on Queen Victoria Street in the City of London, 1939. Note the agility of the fireman stretching to gain the windowsill from the head of his ladder. (© LFB photos.com)

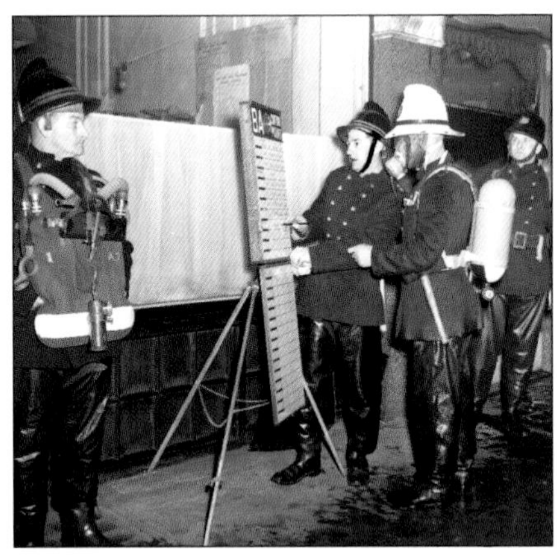

A BA Entry Control during the early 1970s. Note the 'ready for anything' posture of the man on the left, wearing the Proto oxygen set. Also, the white-helmeted officer is rigged in compressed air equipment, quite rare at that time. (© LFB photos.com)

A Pump Escape of the 1970s and the sort of appliance featured within the accounts. This vehicle was known as 'dual purpose', as it could carry the wheeled escape as shown or the normal wooden extension ladder. (© LFB photos.com)

As It Is Sometimes: The huge 1972 Gardiner's Corner Blaze, which took place in an empty store in Whitechapel in 1972, the year I enrolled with the London Fire Brigade. (© Press Association)

'A Night to Remember': The massive fire in Docklands near Tower Bridge, 1973. (© Press Association)

Utter Carnage: The Moorgate disaster, 1975. Crews working inside a wrecked carriage. (© Press Association)

Moorgate, 1975: A soot-covered survivor at street level. (© Press Association)

Moorgate, 1975: Confined-space rescue efforts in the 'tunnel of death', as the press named it. (© Press Association)

Tooley Street: A massive Thames-side fire in a disused cold store, early 1970s. A very smoky job. (© Press Association)

INSET: BA crews awaiting their instructions to enter the smoke-filled warehouse at Tooley Street. (© Press Association)

Weary, blackened and sweaty BA men back in the fresh air and safety of the street after battling the Tooley Street blaze. (© Press Association)

'Little children who suffer, in the sense of early hours. God give us strength to save them. Stops the Devil picking flowers.'
(copyright unknown)

Chapter 15

The Royal 'A'

It's a little after 4 a.m. in late October and our Pump and crew are in a street just off Shaftesbury Avenue, on Soho's ground, in the bright lights of the West End. This street, in the heart of Theatreland, is the London equivalent of New York's Broadway; we joke that some of the fire crews around here think they are film stars, as every time a documentary is made on the fire service, A24 Soho, or more usually Divisional HQ A21 Paddington, feature in it – hence the teasing tag A21 'Hollywood'.

We're here attending a huge blaze in an office-cum-advertising agency's block and have been at the scene since 1 a.m. relieving the first attendance crews and are due our own relief shortly. Coincidentally, Carmel and I will be back in this neck of the woods in a couple of days' time, when we will celebrate her birthday with a meal followed by a show that, unknown to her, I pre-booked some weeks ago.

The blaze had begun about 9 p.m. last night and had required ten Pumps and a Turntable Ladder to bring it under control. Fortunately, no persons had been involved, but it had been a severe fire and the internal damage is such that the job cannot be closed down until all vestiges of fire have been extinguished.

We call this 'turning over and damping down'. Ceilings, walls and internal partitions need to be stripped of any smouldering material and gently drenched and the accumulated debris dumped in rubbish bins and taken outside, where they are soaked again just to be sure.

'Didn't think we'd end up in the "Royal A" tonight, fellas,' quips Paddy Mulligan.

The 'A' Division covered the West End and immediate surroundings. Within its boundaries are the Royal Palaces and the Royal Boroughs, hence its nickname – one that gives a lot of pride to firemen and officers of the division but created a lot of ribbing from the firemen from the other ten spread across the capital. It is a good-natured mickey-taking, but it masks a deeper rivalry relating to how the best Station Officers get their own men on the individual watches to believe they are as good as, if not better than, any of the others.

Inter-divisional technical quizzes, Pump competitions and sporting contests all play their part in this pursuit of excellence. The end result is, in most cases (for there are bad apples in every barrel), an efficient fire force. If every fire station works to be the best, then when the major emergencies occur and crews have to work together from stations far distant from the incident scene, the public is assured of the highest standard of operational performance.

* * *

'Time to get up!' Carmel's voice comes through my restless sleep. Although tired from the West End relief, I seldom sleep well either at the station or in between nights.

It's just after 3.30 p.m. and I have to leave at 4.45 to be sure of not being too snarled up in traffic and thus late for our second night duty. Just before leaving, I hold Carmel close, but I know that it is far harder for her and the other wives, partners and girlfriends than for us firemen. Even when relatives or friends and workmates are around to help temporarily lessen the worries she feels, the truth is that on the long night tour she finds things hard.

It only takes a radio or TV news flash reporting a major blaze for the real tension to begin. The sound of a car door closing in the early hours after seeing or hearing such a news flash, then the sound of footsteps on the street can bring on the thought of a

Senior Officer arriving with bad news. When no knock sounds or doorbell chimes, she will lie there with stomach churning and heart racing, chastising herself for having such a fertile imagination, but she will say it is the price paid for loving a fireman.

As I pull away, Carmel can be seen in my mirror, waving from the front door, and into my mind comes the placard fastened to the mess room wall of the South Bronx firehouse, which I remember from my visit to the Fire Department of New York a short time back.

The Lambeth Headquarters of the London Fire Brigade on the Albert Embankment houses a memorial to all its members who have died in times of peace and war. In the great uncertainty of emergency rescue and firefighting, I was often reminded of those sobering words, 'Tonight could be the night', but you cannot dwell on such morbid thoughts too long. If you did you wouldn't ever go on duty. In any event, danger shared with others is always going to be less of a burden than danger faced alone. Although firemen receive medals and commendations for valour, their individual citations cannot be achieved without the heavy support of their crewmates. It is a team occupation and within that team, bonded by the joint facing of peril, men find it easier sitting or resting, awaiting the toll of the bells, than do their loved ones back home, and it is all too easy to forget this as we enjoy the camaraderie and banter of the fire station.

'What's on the menu tonight, Jerry?' asks Ricky Tewin, as we sit at the table in the long, narrow mess room sipping mugs of tea before getting on with the evening's routines.

'Bangers and mash lubricated by my delicious onion gravy, followed by tinned mandarin oranges and condensed milk. A meal fit for a prince,' replies Jerry Pashleigh, who is 'chef' on this night.

'If it's fit for a prince, you could have it sent over to the Royal A Division where we were last night, seeing that most of their guys think they're royalty,' I say, but only in jest. I had several mates who were serving in that central district to where most of the capital's tourists flocked and I would have entrusted them with my life.

'Princes? More like Ponces! Send the fruits of my slaving over a hot stove to those show-offs – what do you take me for?' jokes Jerry. 'Anyway, unlike them, we've got the Tower of London on our patch, so they'd better watch their step as we can get them locked up for treason.'

It's 6.20 p.m. Before sitting down, we were all on roll call, fully booted and spurred. Sub Officer Jack Hobbes had checked that everyone who was meant to be there was present, and detailed who was riding the Pump and the Pump Escape, who were the drivers, who were down as BA wearers and who was duty man for the shift.

The duty man goes to the teleprinter when it sounds to receive the emergency or administrative communication. He also sources the shortest and quickest route to the address received from an indexed box of route cards. Such is the Brigade's tradition of rapid response to all emergency calls that the duty man has to sleep in a special curtained-off cubicle so that he can have the orders and route cards ready, no matter how early the hour or how bleary from a busy shift he is.

In addition, he has to report to any visiting Senior Officers when he hears one ring of the call bell, which signals that officer's presence. One of his further key duties is keeping an accurate station log, which is hand-written and records all personnel on duty, the appliances available, who is in charge of them and who has been detailed to wear BA. If a man is relieved by an oncoming shift fireman and the duty officer in charge agrees the relief, the duty man must clearly annotate in the log this rider change; the nominal roll board kept on each appliance is annotated also.

After the roll call, drivers check their appliance's fuel level, and satisfactory operation of all lights and warning devices. The other firemen check every locker to ensure that no item of gear is missing and make a similar check of ladders and hose reels. The nominated BA men check over their sets and enter their name and oxygen cylinder contents on a plastic tally with a greasy pencil.

It is all about preparedness; a lot of the thinking behind the Brigade's policies stemmed from the 'Ready Aye Ready' traditions of the Navy, whose former seamen once comprised most of the Brigade's establishment. No use waiting for the storm on the high seas to burst before all hatches are battened and ropes and lines tested. Such a sloppy approach could result in a ship's going down with all hands.

By the same yardstick, it is no use waiting until we get to the emergency to find that a vital piece of equipment is missing or that a BA cylinder is out of oxygen – these could also result in a life-loss disaster, including the lives of firemen.

So once all of these checks have been satisfactorily completed, the duty man is informed and he enters the results in the log book; should a fatality occur, this will be legally required as evidence in any investigation.

We sit down to eat dinner and as usual the conversation turns to controversial topics. I bring up one of my pet irritations: why it is taking the Brigade so long to replace the oxygen BA with compressed air sets.

'The Proto set is a great bit of kit, guv,' I say to Biff Sands as we take the first gulp of tea, 'but when are the Brigade going to catch up with our mates in most other parts of the UK and get compressed air sets? If New York City firemen can trust them, surely it's high time we did, too.'

At the time I transferred to the capital, my former brigade had been using the full face mask compressed air (CA) sets for some years. The Proto oxygen equipment (its mouthpiece and nose clips always reminded me of the French scuba diver Jacques Cousteau, whom I had watched on the TV as a teenager) had been rendered obsolete in the late 1960s by my former brigade, along with many others in the provinces. Not only was the CA set lighter, but its full face mask allowed men to orally communicate, rather than having to rely on grunts and hand signals to communicate, which wearing a mouthpiece necessitated. A full face mask did exist for the oxygen set, but those were confined to the 'Emergency Tenders' for use in corrosive atmospheres such as are found in refrigerated warehouses in which the refrigerant has leaked.

Biff Sands gave a slightly weary look, showing that he had heard me raise my concern more than once before. 'Well, as you know, we have some CA already, but there are a couple of reasons why we have no more at this stage,' he went on quietly. 'The first is the cost of re-equipping all of the capital's 114 fire stations, and the second is to do with the duration of the air cylinders compared with the oxygen set.

'The Proto set is bulky and takes a long time to clean and test before it can be reused. But it has almost double the 30 minutes duration of the CA set and in some of the deep basements and underground rail tunnels, something that few other brigades have, that extra half-hour could be a life saver.'

'It's not only that though, guv, is it?' Dick Friedland chimed in, pulling on his pipe. Through a veil of smoke from the rich 'baccy', he continued, 'What the CA set can't give you that the oxygen set can is the ability to extend the supply if a man is lost or trapped. As long as a bloke's conscious and can operate the valves, it can be doubled or trebled if you lie still.'

'Which gives all that extra time for a rescue to be made,' Biff adds.

We are getting into a heated debate about the merits and demerits of the CA and oxygen kit, including the argument that pure oxygen is a hazard in a flashover, as it can intensify burning, when the bells toll, ordering the Pump to '*Person locked out – danger of fire*'.

'Danger of fire?' says Jack Hobbes, as he comes back into the mess. 'The public have got wise as to how to save themselves the cost of a locksmith!' he adds with a sneer, shaking his head. 'They know if they tell the operator they think they've left a pan on the stove, we'll have to turn out in case there's a real risk of fire – it's about time we started to bill the crafty buggers!'

By the time the Pump returns, it is time to sample Jerry's bangers and mash. I slice the juicy brown pork sausage and sink my teeth into its hot centre just as the bells toll again.

'Both!' yells Ricky Tewin, duty man from the watch room door, as we all plunge down the pole and see both red and green bulbs glowing.

'What we got, guv?' I shout above the throaty roar of the diesel engine, as the Pump clears the station just behind the Pump Escape.

'*Smoke Issuing. Multiple 999 calls, general warehouse*,' he shouts back above the two tones, which his boot is operating in short bursts.

The street where we are to find the fire is as flammable as a ton of petrol-soaked firelighters, as Dick Friedland mentions, subconsciously tightening the press stud another notch on his helmet strap, sensing that we have a working job. It is 8.10 p.m. and we are a good quarter of a mile from the address given.

'Bugger me! Look at the smoke! It's like the pea-souper fogs we used to get back in the 1950s,' exclaims Dick, as the Pump slows to a crawl amid the rolling waves of heavy brown smoke which has reduced visibility to a few feet.

If the smoke is so dense 400 yards away from the address on the call slip, then we look to be in for a hard night.

For a brief second, as I smell that so familiar pungent, acrid odour, my stomach churns with apprehension. I recall the hug I gave Carmel five hours back – but it's no use dwelling on copping your lot. Shut out any morbid thoughts and concentrate on doing the job you chose of your own free will, I urge myself. If I had wanted it quiet and relatively safe, I could have transferred to a station in the deepest rural countryside, not to the centre of the Big Smoke, an extremely apt title, I think, as we grind to a halt opposite a cobbled yard veiled in thick smoke around which several five-storey warehouses are crammed.

'Seeing it's almost Bonfire Night I think it might be a sky rocket firework burning on the roof, officer,' a tall, 'theatrical type' woman in a long black coat is saying to Biff Sands.

'Firework, madam? It must be a damn big one going by this smoke. Do you know what the warehouse contains?' Biff enquires, his eyes rapidly scanning the scene and his mind assessing the size of the fire and the prospect of its spread as he speaks.

'Oh, all sorts of stuff, a general repository for everything including the kitchen sink – but will you be able to put the fire out quickly, as a lot of us on this side of the courtyard have studios and art galleries that are our livelihoods,' she rattles on anxiously in a cut-glass gin-and-tonic voice.

'We will do what we have to do, madam, but this is a serious blaze and I must ask you to leave the area at once,' Biff replies emphatically, then he turns to his crew. 'Pitch the extension ladder to these buildings opposite. I want to have a closer look higher up towards the roof of the fire building,' he instructs and in a flash we have the 30-foot wooden ladder pitched. For all his height and bulk, Biff ascends the ladder like a nine stoner, the extensions bowing with his weight.

He can see that heavy smoke is venting from the roof of the five-storey warehouse. 'Sky Rocket?' he mutters to himself. 'There will only be one rocket and that will be from the Divisional Commander if I don't get some reinforcements here quickly. This place is going like an express train.'

'Jerry!' he shouts from his lofty perch. 'Send a priority: *Make Pumps Eight*.'

Bang go those bangers and mash, I think. The radio message will see five more pumps and several chiefs en route in seconds. This looks like being a protracted job and, as if to confirm it, the viscous smoke racing up from all of the warehouse windows on the top two floors is sliced apart by vicious long tongues of red and yellow flame. To use a wartime term, the warehouse is 'well alight'.

Within only a minute of the first assistance message, Biff sends a further one, calling for the Pumps to be made up to ten, plus a Turntable Ladder to protect the exposed warehouses adjacent.

Once eight or more Pumps attend fires, the Emergency Rescue Tender and Control Unit are automatically dispatched.

Biff's informative message tells its own story: '*General warehouse of five floors, 400 feet by 100 feet. Whole of third and fourth floors, and whole of roof alight. Remainder of building smoke-logged. Four jets, in use*.'

Within ten minutes of that message being transmitted, the role of our Station Officer within the command structure changes from his being in overall charge to being a working hand because a more senior officer has arrived and taken over the command.

I am sure that Biff Sands, for all his leadership and man-management abilities, enjoys sitting with his men astride a parapet wall 50 feet up and directly opposite the blazing warehouse. He enjoys it because, for a time, he is able to recall what it was like to be a young fireman, free of the extra responsibilities and accountabilities that promotion brings. Holding a large nozzle that is delivering several hundred gallons a minute into the inferno, which we later learn is the result of an arsonist, reminds him how much he enjoys the raw essence of being an inner London fireman. He is able to enjoy doing what, in essence, he joined the brigade to do – get into the thick of things – and like all of those dedicated firemen of the period, he does just that, satisfying that inner pleasure derived from quelling fierce flames with the powerful jet.

We are relieved a few hours later, returning to the station, where we re-stock the pumps and put on clean, dry gear, ready for the next shout, which we will respond to with that high sense of confidence that comes from being led into the fray by a real guvnor like Biff Sands.

Chapter 16

False Alarms and 'Red Biddy'

We have just got back from the fourth malicious false alarm since coming on at 6 p.m. Every time we respond to this type of call, the public is put in danger, because should a genuine emergency arise there are a reduced number of Pumps and crew to respond.

Pumps have to be dispatched from the nearest station, which, at busy periods, could be fifteen to twenty minutes away, rather than the five or ten minutes from the local station. Those lost minutes can be crucial to the saving of life. In addition, where a real emergency has occurred and the fire is spreading rapidly, it can get away to the extent that a building can be lost or far more badly damaged than it would have been had we not been otherwise detained on a wild goose chase, searching for a fire that did not exist.

About a third of all calls received are false alarms. Not all are made with malicious intent: often the person making the call genuinely believes there is a fire in progress. In other instances, the call-out has derived from defective or over-sensitive fire-detection systems or through fluctuations in the pressure within the water mains, feeding sprinkler installation alarms.

'What sort of no-brain dim-wit thinks it's a hoot to put in a 999 call just for the fun of it?' exclaims Paddy Mulligan, stand-in 'chef' for the tour, as he slices the big block of corned beef that he will make into a hash with a mashed potato side dish.

'A no-brain dim-wit, that's who, Paddy,' Ricky responds with a sarcastic look.

'Yeah, OK, Jerry. Let me elaborate,' replies Paddy, as he puts the corned beef into a large shallow pan atop the six-ring gas stove in the corner of the long galley kitchen. 'What I was trying to get to is what sort of brain is inside the skull of whoever gets a kick out of making false alarms?'

'The brain of a no-brain dim-wit, as I just told you!' grins Jerry, to the accompanying laughter of the three others huddled around the cooker.

'What the bloody hell are you two going on about? You sound like two fifth-rate comedians in the pub over the road,' chimes in Sub Officer Jack Hobbes, who has come into the mess room, connected to the kitchen via a large serving hatch.

'What do you expect, Sub, from two no-brain dim-wits,' adds Lofty, with a side-to-side shake of his close-cropped head.

'OK, guys. Be kind to me if you want my delicious hash this side of midnight,' Paddy retorts with a wide grin.

'Looks to me, Paddy, that you're making a real hash of it already. Are those two-foot-square blocks of corned beef your interpretation of the cooking term "diced"? Jesus, is that how you make it in the Emerald Isle? They look like the pile of paving slabs Murphy Civil Engineers are laying in the High Street!' says Lofty, with mock disbelief.

Paddy, his red Irish face wearing a wide grin, is about to mouth some response when the bells toll.

Paddy could have adopted a serious stance from the outset when the topic of hoax calls was raised, instead of going along with the good-natured joshing that is a regular part of fire station banter; he could have easily ignored being ribbed about his cooking and concentrated on the annoying and rising incidence of false alarms. His cousin, who served with New York City's fire department, had told him of the men who had been fatally injured by being thrown from the back step of the Pump whilst speeding to a malicious false alarm, and there is always the same hazard of an accident en route when we are out on these criminally reckless wild goose chases.

As we hustle across the floor to the pole drop, we'll all be thinking, is this going to be the fifth mickey of the night, or is it the real thing? That is the eternal problem posed by false alarms. We never know until we arrive at the scene whether the shout is for real or yet another hoax. The day we play cry wolf and dawdle to the address is the day it will be a raging inferno with people screaming for rescue from all floors.

There is screaming as we arrive in a seedy, run-down backstreet of a dilapidated, unloved East End district. But it isn't coming from all floors. It is from the mouth of a man who is alight from head to foot, looking like a Roman candle on Bonfire Night. He is reeling from side to side on the front path of a derelict three-storey house from which brown smoke clouds are billowing menacingly.

'Get the hose reel tubing on him now!' shouts Biff Sands. 'Jack, get yourself round the back for a look-see. And Jerry, send a "*Make Pumps Four – Persons Reported. Plus extra ambulance required,*"' Biff shouts to Jerry Pashleigh, the Pump's driver, holding up four of his meaty fingers to reinforce the point.

The street we've been called to has been the site of more than a few nasty fires and fatalities involving tramps and winos over the years. The usual scenario is that they get a fire going either in the debris-laden rear yard or in a fireplace inside the rat-infested houses.

Their tipple is virtually anything with alcohol in it, but a potent mix of throat-searing methylated spirits and cheap rotgut red wine, a concoction known as 'Red Biddy', seemed to be the favoured route to a drunken stupor and oblivion.

I have yanked the 'tubing' off almost before Jerry has got the handbrake on and, as soon as he engages the Pump's power take-off, I am dousing the burning man with the powerful conical spray. As the flames disappear, I can see the terrible skin-splitting burns to his face and hands, with the surface almost bubbling, and his piercing screams of agony are filling the air.

While I am doing this, Ricky Tewin and Paddy Mulligan have laid out a line of inch and three-quarter hose and Jerry, after sending the assistance message, has connected the Pump up to a hydrant.

As soon as they have water, they go in. They keep low, seeking out the source of the blaze and searching as best they can in the heavy smoke, which has banked down to the floor, darkening the rubble-strewn hallway further with its vile lung-damaging thickness.

Toby Smollett has got a burns sheet from the first-aid box and, as he tries to cover the vagrant's horrific upper body burns, I follow Ricky and Paddy, who are working their way into the front hallway of the house. We are all suffering a lot from the acrid brown smoke.

'Looks like it's at the end of the hall on the left,' coughs Paddy. There is a dancing ruddy reflection illuminating the underside of the gravy-thick smoke that is painful to the nose and throat.

We haven't been working the line into the hallway for more than a minute when Paddy yells, 'There's a body here on my right!' He has stumbled into the unconscious form as he worked the line forward.

'I'll try to drag him out,' I splutter, as more of the fiery hot smoke goes down my throat, feeling like a big spoonful of Tabasco sauce.

Stumbling on my knees, I feel the soft shape of a body. I can just make out from the light of the flames, which are now licking menacingly along the underside of the ceiling above us, that it is a man. I get my hands under his armpits. He is fortunately in a position where his head is facing the house entrance. He is only a slip of a man, so I easily drag him, his feet sliding over the rubble, to the refreshing coolness of the clean air in the street.

Both the ambulance and the reinforcing Pumps have arrived. Two attendants are loading the hideously burned man onto a trolley.

When they see me, they stop. One attendant dashes over and drops onto his knees beside the man. He puts his head on one side, listening and feeling for signs of breathing, then feels for a pulse in the victim's neck from which the seared skin is peeled and hanging.

'He's still alive. We'll take him as well. Might be too late by the time we get a second ambulance here,' the attendant sings out. A couple of police constables who have arrived on foot give a hand in the loading of the men into the ambulance.

BA wearers from the reinforcing Pumps have started up and go in to search all floors after relieving Ricky and Paddy, who, not being protected by BA, must be just about all in by now. We have only been here ten or so minutes, but it feels much longer.

I hear the retching coughing of Paddy and Ricky as they stagger back into the street. Paddy's normally pink Irish face is now like beetroot. He leans on outstretched arms on the side of the Pump, his upper body parallel to the hard cobbles of the street, onto which he is violently sick, the poisonous pungent fumes having got to him.

Ricky drops to his knees and Barry dashes over with the black-and-white spare oxygen cylinder and cracks the valve in Ricky's very flushed face, which glistens with perspiration under the street lamp. But Ricky is a lot younger than Paddy and recuperates much more quickly, as only those on the right side of 35 can. He stands up and goes over to Paddy, offering him a recuperating whiff of pure oxygen, having caught a glimpse of him opening one of the Pump's delivery valves and filling his helmet with cold water, pouring the lot over his steaming hot head, his black tousled hair flattening to his skull.

'Bloody hell, Paddy, mate. If you can be spewing up like that for a bit of smoke, you'll be in a right state if you get a bellyful of that concrete aggregate you call a corned beef hash,' chides Ricky with a shake of his head. 'Joking apart, are you OK, mate? It was a real bit of the old thick stuff in there, wasn't it?' He then spits a thick ball of soot-stained mucus onto the street.

'Don't you worry about it, young Ricky. And as for my supper, you have no worries over that at all! Gimme another lungful of oxygen,' he replies, slumping onto the side step of the Pump Escape, wiping the vomit from his chin with the sleeve of his tunic.

ADO Ernie Tappison has come on and been briefed by Biff before they enter the fire-ravaged hovel together. The blaze had started in a rear room, no doubt a consequence of the winos' constant attempts to have a warming fire on the go. Fortunately, the two victims were the only ones present or casualties could have been a lot heavier. It is likely that, stupefied by the meths and wine mix, the badly burned vagrant had got too close to the fire. We have attended more than a few of these types of jobs in these derelict premises within the Borough of Stepney. Several of these 'meths men' had paid the ultimate price and sadly the two at this fire did so also, both of them dying in terrible agony over the following days.

The ADO congratulated all present on their actions before Biff instructed us to make up the Pump Escape's gear and get back on the run, ready for whatever else the night might throw at us, be it another mickey or a drama like the one we had just handled.

The Pump remains for another half-hour or so, turning over and damping down.

Some return to resurrect Paddy's hash, or for the less adventurous a call-in at the chippy – a much more conventional hunger cure.

However, even after our usual antidote of black humour round the mess table, not everyone is feeling much like eating following the night's tragic events.

* * *

As I have said, a good 30 per cent of all responses made were to various types of false alarms. The following are provided to give a flavour of the diversity of these:

A Troubled Mind

She is a well-built woman of about 5 ft 4 in. and ten stone, and is clad in only a small pair of white pants, which make the skin colour and raven hair of her Asian origins appear even darker. Both hands are firmly clasped around a long-bladed carving knife, which she is holding in front of her naked heavy breasts.

We have been called to a fire in the kitchen and on arrival were met by this woman, wild-eyed and screaming at a small, balding and bespectacled Indian man to 'Get out, get out and leave me alone!'

A thin blue haze of smoke hangs in the air of the kitchen, which is accessed directly from the street and to which the door is ajar. A strong smell of curry and other Asian spices mingles with the thin smoke.

'Send a priority message requesting urgent attendance of police,' says Station Officer Tuke, without taking his eyes off the knife-brandishing woman for even a second. Gingerly, he moves into the kitchen.

'We've been called to a fire here and I would like to check if everything is OK,' he enquires quietly of the man. 'Are you able to tell me if the 999 call came from this address please?'

The small, slight Indian eyes the woman warily, as if too afraid to answer.

'I want him out of here!' she screams and at the corner of her lips white saliva can be seen, the sort of fluid that sometimes accompanies a violent rage.

'Please, please, Sunita, there is no need for this. Let me come by to talk to the fireman, please,' he pleads.

'No, no, no! Keep away or I will cut you in half,' she yells, her black eyes glinting.

Whatever the reason for this scene, the woman certainly appears dangerous. The look in her eyes, the lethal kitchen knife, her half-naked appearance and her loud threats suggests either a grievous wrong carried out by the man or a woman whose mental equilibrium is in serious doubt. Station Officer Tuke manages to take a quick look at the gas cooker and, although there are a couple of pans on the rings, there is no sign of overheating. But there is definitely a haze of smoke and a smell of burning in the air.

Ben Tuke knows it is too risky to force his way any further into the house at this stage. There is no obvious sign of fire, but a proper check will have to be made before the scene can be left. At the moment, though, it is simply too fraught. As he contemplates his next move, two policemen appear at the front door. One is shortish, about 5 ft 8 in. with carrot-red hair, the other well over 6 ft, balding, with the thick paunch redolent of a man who likes his beer.

'What's the score, guv?' the redhead enquires.

'No idea,' the Station Officer replies. 'We got a call to a fire at this address. There's a bit of smoke and a burning smell but nothing evident. It looks like a domestic squabble.'

In spite of the seeming danger in the situation, the small red-haired constable goes straight into the front room, where the woman stands brandishing the knife, still staring at the small man. 'Right, listen to me,' he says in a hard, uncompromising tone. 'Put that knife down on the floor now or I will take it from you, do you understand me?'

'I want him out of this house. He's a filthy dog!' she spits out, but already the venom in her voice is diluting. Perhaps the sight of the policeman and the tone of his voice is jolting her mind back to a more rational state.

'Just put the knife down – now!' comes the still insistent voice of the policeman, his freckled hand moving onto the top of his baton, ready to draw it if she tries to use the carving knife.

'Please do what the officer says,' pleads the little Indian. 'You know we can sort this mess out.'

Something must have registered in the agitated mind of the woman, as she suddenly throws the knife down onto the carpet and begins to wail and sob hysterically. The tall policeman places his huge hands on her shoulders and leads her to a battered sofa under a window that is shielded from the street by some shabby net curtains.

'Now, tell me, what's all this about?' he asks.

As she begins to talk, the slight form of the Indian man moves towards the Station Officer. 'I am so very, very sorry. Please let me talk to you in the kitchen,' he says, his eyes still watching the woman as he moves towards the cramped, seedy kitchen. 'She is sick, you see. She has schizophrenia and has to have an injection every four weeks to control the worst symptoms. Many times when there are only a few days before the treatment, she gets like this and twists something trivial into a major crisis. No matter what I say or do, she simply does not believe anything at such times.

'I became very scared when she started to become violent earlier, over nothing at all. I rang the police, but they were taking too long, so all I could think of was calling you out. I don't know why I did, but I was so desperate that something made me call the fire brigade.'

'So there has not been a fire, then?'

'Oh no, there has been no fire.'

'But the slight smoke and the smell of burning – where is that from?'

'Oh, that is the oil and scent we light and burn as one of our religious rituals. Will I get into trouble, sir?'

Station Officer Tuke looks the small Indian in the eye before saying, 'I am not happy that you called, saying there was a fire when you knew there wasn't any such thing. Someone could have died if a genuine incident had occurred whilst we were here, but I guess you have had a big shock, seeing your wife getting so angry and threatening you with that knife, and you panicked because of the police delay in getting here. Taking that into account, we'll leave it at that, sir, and I hope that things can improve for your wife.'

'Thank you a million times,' says the grateful man. He turns to go back into the lounge to speak to the policemen and his troubled wife.

We mount the appliances from the pavement, where we had waited quietly until this 'crisis' was over, then book ourselves available and ready for whatever else the people and properties of London might throw at us.

A Disgruntled Keyholder

It is just after 2 a.m. when the bells toll, followed by the cryptic printed message: '*Automatic Fire Alarm Actuating [AFA], Eastern Textile Warehouse*'. The call is on our ground. We are on scene in a few minutes and the hee-haw of the third Pump's horns can already be heard a good mile away.

The warehouse is a four-storey redbrick structure about 250 feet long by 60 feet wide and is crammed with a wide range of cloth from across the world. It is a place to which annual operational familiarisation visits are made by all watches from our fire station, so we have a good idea of the building's layout.

Other than the shrill ringing of an external bell, immediately apparent on dismounting, everything seems as still as the black ebony surface of the nearby canal.

'Sub, you take your crew around the back, we'll check the front!' the guvnor's voice rasps out in the still air. 'And send an informative,

John: "*AFA actuating, no sign of fire, search of exterior being made.*"' John Lydus, the Pump's driver for the shift, repeats the message in the time-honoured fashion of the Navy, confirming he has got the content right. He picks up the handset to transmit just as the blue beacons of the third Pump come into view from the junction 100 yards ahead.

We check the front of the warehouse, our noses twitching like hunting dogs, as we try to detect that unmistakeable whiff of burning. We place the back of our hands on window panes to check for heat and shine our lamps through windows, trying to catch in the light the reflection of smoke – all the time, the shrill bell trings away.

The other crew do an identical exercise at the rear, but they are able to climb an external fire escape and check conditions at the fire exit door on the top floor, where the smoke and heat of any lower fire would be most likely to have risen. No indications of fire are evident.

This isn't the first time that we have received AFA actuating calls to these premises in the early hours. In fact, the station's log shows the other watches had attended as well, perhaps three or four times over the last year. These calls, added to the brigade's overall statistics, record that some 30-plus per cent of all calls received are to AFA alarms.

After about ten minutes, the Station Officer asks John to verify the origin of the AFA actuating and within a few seconds Fire Control come back to say that it has been routed by a Central Alarm Station company and that a 'fire' condition is still showing on their indication equipment.

The officer in charge is now faced with something of a dilemma. The whole point of an automatic fire alarm is just that: to automatically detect and alert of a fire situation. It is the relative cost of such installations, over time, compared to the wages of night security staff, that first made the systems attractive to small to medium companies, especially as insurers will offer a reduction in premium if an approved system is fitted.

So, a fire alert is showing, but there is no outward manifestation of fire. Increasing attendances by brigades have demonstrated that an alarm can sound and show a fire even when none is present.

Were the Station Officer to base his decisions on these past negative experiences and decide to risk sending a Stop message, stating that the alarm was caused by 'fire alarm system fault', he might well get away with it. But sod's law will dictate that he will come unstuck eventually. A repeat call would be received and on return the premises could be alight, with all the consequences for the business, not to mention the officer in charge, who could face dismissal and the fire authority a possible civil action for loss sustained. As a consequence, the officer in charge has two options.

The first is to force an entry into the premises on the basis that the alarm indicator is showing a fire situation. The second is to request the attendance of a key holder via the police, then wait outside with all three appliances until someone turns up to open the place and allow a top-to-bottom search inside. The Fire Services Act allows members of a fire brigade to lawfully enter premises where there is reason to believe that a fire has broken out, so a forced entry wouldn't be illegal.

As all external checks have not discovered outward indications of fire, the usual route, option two, is taken. A radio message confirms that the warehouse key holder lives some miles to the north of London and it is estimated he will arrive in at least 25 minutes. In due course, a large white Mercedes saloon draws up. The key holder and proprietor, a short, balding man of Middle Eastern appearance, wearing a camelhair coat that nearly touches his mid-calf, gets out.

As he opens the large iron-grilled front entrance door, he remarks to our Station Officer, 'I'm sorry that you have had to come out like this – but why are there so many fire engines, there doesn't seem to be a fire?'

The guvnor towers over the man, his large white-combed helmet making him appear at least another four inches taller than the six-feet-plus he actually is. 'Well, sir,' he replies in a clipped but respectful manner, 'the fact is that your automatic fire alarm is showing a fire situation. This area is what we call "high fire risk" and in those areas we always send three pumps to any alarm of fire within buildings – and that includes alarms from automatic detectors. I have attended these premises at least twice in the last year, sir, and I know other officers have attended here also. Each time it's been a system fault, but we never assume this to be the case.'

'Well, yes, I understand that, officer. Of course, I wouldn't want you having to come here if there is only a fault, but then I have paid a lot of money to have this system installed so that I can have the peace of mind that if a fire breaks out it will be quickly detected,' he responds wearily.

'Yes sir, and it is a wise decision. But perhaps you could have the alarm engineer attend later today to try to establish why you are experiencing these faults. I will contact the local fire safety officer and ask him to arrange a specific inspection with you,' says the Station Officer.

He would have liked to have gone on about the fact that during the best part of an hour of being detained, someone's life could have been lost or property destroyed; he knows the fault is most likely to be some defect in the apparatus or the proprietor neglecting to maintain the appliance, but he is impotent as to taking any sanctions so he has to bite his tongue.

Within a quarter of an hour, we have satisfied ourselves that there is no sign of any fire and the all too familiar 'Stop' is transmitted: '*Alarm caused by AFA*'.

The warehouse proprietor's Mercedes slides away, its silky-smooth purr contrasting with the harsh roar of the diesel engines of the three appliances as they pull away in a cloud of blue diesel smoke.

King's Cross in the Early Hours

The two Pumps have only travelled a few hundred yards when the radio bursts into life, calling up the Pump with a message prefixed with the words 'Priority': '*Order your Pump to Fire, York Way, adjacent King's Cross railway station*,' says the smooth female voice of the Control Operator.

'Received and mobile to fire, York Way, adjacent King's Cross railway station – acknowledged,' replies the Station Officer, simultaneously giving a blast of the siren with the switch under his right boot as we approach the traffic lights, just turning from amber to red.

Drivers of fire brigade vehicles proceeding to emergency calls are permitted to exceed the speed limit, provided all warning devices are operating, but are otherwise subject to the same requirements and regulations that apply to other drivers. Traffic lights are no exception and, legally, should a driver cross a red light and be involved in a collision, then that driver can be charged with a variety of offences. However, given the amount of traffic lights across Greater London, particularly within the inner suburbs and the West End, many minutes would be added to response times if a halt was made at every set and those extra minutes could cost lives.

The normal procedure, therefore, if the lights are at red or about to change is for the driver to ensure that all audible devices are operating, in addition to blue beacons, and to cautiously 'peep and creep' across the junction. On the vast majority of occasions, other motorists will pull over or slow down to allow the appliances to cross, but the danger comes not only from inattentive drivers of other vehicles but also from the elderly, whose faculties might be impaired, and the younger element, who delight in filling their cars with high-volume music. Both situations may mean that sirens are not heard or beacons remain unseen until it is too late.

Given the very high volume of calls responded to by London Fire Brigade each year, the number of serious accidents en route is small, but fatalities have occurred across the country. In a well-publicised case in the 1960s, Lord Chief Justice Denning laid down that no fire was so urgent that risks could be taken in crossing traffic lights at red and that the onus lay squarely with the appliance driver.

It is doubtful if Jerry, our driver, is thinking of that distant legal ruling as he cautiously crosses the red light, but he is very aware that even with the Station Officer sitting next to him, the onus of responsibility, should he have a collision with another vehicle, cyclist or pedestrian, lies solidly at his feet.

Even in the world's busiest fire departments most responses to fires end up as either incipient incidents or one of the categories of false alarms. How many times had we thundered and blared along streets thick with all manner of vehicles, wandering pedestrians and seemingly suicidal cyclists – riders who often appeared from nowhere into the path of the charging red engines, usually oblivious to the eight tons of steel atop them – only to return, slinking back in a silent embarrassment at the cacophony of siren and bell only a few minutes earlier?

When our rapid response enables a successful rescue or a potentially serious fire to be put out before it becomes an inferno, we are grateful for the verve and skill of our drivers. However, had we piled into a car or pedestrian, killing an individual by shooting a red light, where would we be then? How would the headlines read: 'Five members of the public killed in road accident, fire crew members also possibly victims in race to a false alarm'? Perhaps such reasoning was in the eminent judge's mind when he made that historic judgment.

The electric blue flashes of the beacons become momentarily more vivid as we hit the relative dinginess of the long parapet walls above the Copenhagen rail tunnel on York Way, just as we hear over the radio '*Booking in attendance*' in a message to Fire Control from the local stations.

King's Cross station was a good place to go after a day responding to fires and other incidents. Often two or three of us could be found in the bar, sampling a few pints of the more tasteful beers on sale. Our thirsts adequately slaked and with the appetite whetted by the best beers, we would often indulge in a fish-and-chip supper in a greasy spoon café alongside the Pentonville Road. Or we would jump on the underground and go 'Up West' for a few more drinks in the multitude of bars around the theatre district.

'I could kill for a juicy T-bone steak and a bottle of red,' whines Jim Peating, as the Pump brakes with a squeal outside a small shop that sells model railways. We park opposite the two local station Pumps, the combined effects of our blue rotating beacons strobing the dingy buildings and attracting a gaggle of creatures of the night from the station's frontage.

'You'll get more than a bottle of red around here, Jim,' the guvnor chuckles, pointing to a bevy of ladies of the night, teetering in their stiletto heels and tight leather miniskirts. The area around the station, and in several of the adjoining streets, is an infamous red-light district and like a number of other parts of the capital there has been a growing presence on those seedy streets of drug pushers and users.

Only a few weeks earlier we had attended a good working job in the small hours, in a nearby squat occupied by an assorted crowd of glue sniffers, heroin addicts and prostitutes. It was a very smoky fire involving mattresses and filthy bedding, and three addicts had been dragged unconscious down a tall narrow staircase. But it was probably the drugs that had initially knocked them out before that brown pungent smoke enveloped them. There is no smoke to be seen here, though, so the officer in charge verifies from Control where the 999 call has come from. All that comes back is that someone had seen a fire near the King's Cross station.

A blare of a two-tone horn heralds the arrival of the Turntable Ladder, an automatic ordering to all calls to 'fire' in this high-risk locality. The indicator plate beside the cab door shows that it has come from Euston. Although we all suspect that this is yet another mickey, chances can never be taken with lives or property, so the officer in charge instructs the Leading Fireman in charge of Euston's ladders to get aloft and have a good look around over the rooftops for any smoke or flame. This is a quite standard procedure, akin to the use of fire lookout towers in North American forests.

Within five minutes, the ladders have been pitched, their steel pawls clanking as the telescopic extensions shoot up to almost 100 feet. The Leading Hand, secured on his platform at the head by a safety belt, scours the buildings around York Way, Pentonville Road and King's Cross Road, as well as the railway arched thoroughfares of St Pancras and Midland Road. The only smoke to be seen, however, is that of the blue clouds punched out of the exhaust stacks of the diesel locomotives as they inch away with their long train of northbound coaches.

'Looks like we've been effed about again,' says the guvnor, with a note of disgust and resignation in his voice. A resignation born out of having stopped trying to analyse the moronic behaviour of whoever gets a perverse thrill from a malicious call to the fire brigade.

'Do you think that some perverted and twisted bastard sees us fly past to a call and decides it will be a wheeze to get us out again by making another mickey?' asks driver John Lydus of the three of us in the rear of the Pump as we head back to the station.

'Could be, John,' replies Jim Peating. 'Psychologists have said that there is a link in the minds of some perverted individuals with everything connected with fire. It's said that they get a buzz watching fire engines racing through the streets, or from standing out of sight in a dark and dingy doorway watching firefighters rigged in their helmets and other gear.'

'Yeah, who's that watching us from the door of the model railway shop?' quips John, and simultaneously all heads whip round to see nothing but an empty doorway.

'Got you!' says John, his grinning face looking yellow in the light from the street lamp. 'In truth, though, I think mickeys are often down to young kids who get a buzz out of mischief.'

Whatever the motives, we all agree that where other forms of false alarms have to be accepted as genuine processes of good intent or the malfunctions of otherwise extremely beneficial detection and fire warning devices, malicious alarms are not only sick but also criminal acts that could cost the lives of the public and of firefighters.

And there was every sign at the time that such calls were increasing rather than decreasing in number.

Chapter 17

Down-and-Outs

Most large towns and cities have their vagrants and homeless, and those who choose to squat in derelict or unoccupied houses, in factories and even inside old cars or vans. London is no exception, and the East End and parts of the inner suburbs have always had their quota of the dispossessed. I only have to look at a large-scale A-to-Z of the East End and read the street names for so many memories to come flooding back into my head; the same head that wore a helmet all shiny ebony-black at the start of a night duty and so often filthy on my return from a smoky, water-drenched derelict house or factory in which down-and-outs lived during the darkest hours.

Brick Lane, Brushfield Street, Fashion Street, Frying Pan Alley, to name just a few, had seen the blue-flashing beacons and gleaming red Pumps more often than any other thoroughfare in this atmospheric locality.

Within such areas, there are usually a fair proportion of buildings in various stages of dereliction. These are awaiting either the steel ball and sledgehammers of a demolition crew or purchase by some developer hoping to carry out a minimal-cost refurbishment prior to making a healthy profit on a future sale.

During the long period I spent working in and around many of the inner London boroughs, calls to derelict premises were a regular occurrence. Although it was easy to treat such fires with complacency, one could never be certain that within the depths of what appeared to be a totally uninhabited property there were not vagrants or squatters sheltering from the elements, often anaesthetising themselves with a variety of alcoholic beverages.

* * *

Our evening meal of boiled beef and carrots has been interrupted at just after 8.15 p.m. on an icy December evening by a call to '*Rubbish alight – rear of Christ Church, Spitalfields*'. This is directly opposite the vegetable market and in the heart of the East End. Such calls within the inner city usually involve a pile of commercial waste, such as empty cardboard boxes, wrappers and refuse sacks, stacked at the rear of a premises and ignited by children playing with matches or by a discarded cigarette end. In other cases they involve a small bonfire set up by some of the area's down-and-outs in a bid to warm their hands and bodies. The location for this call-out is a regular one. The rear of the church is known as 'Itchy Park', its title derived from the actions of those who have worn the same garments for a long time and who have not bathed or showered, in many cases, for months.

The ruddy orange and yellow flames illuminate the weathered faces of about six men and women – faces that are blackened by the greasy smoke of their fire; faces whose deeply etched lines tell of the highs and lows of life, reminding me of contour lines on a map, indicating the existence of high and low terrain.

A passer-by had noticed the flames. If unfamiliar with the area, and with the vagrants' haunt of Itchy Park, you could indeed easily construe the flames as a sign that a more serious blaze is afoot and consider it prudent, in this high fire-risk locality, to dial 999.

A virtually toothless woman, her head half-bound in a ragged tartan scarf against the cold, and grasping a wine bottle of which all but the neck is inside a brown paper bag, is sitting leaning against a wall. Her croaky voice is giving a faltering rendition of the nursery rhyme, 'Oranges and lemons say the bells of St Clement's'. She breaks off to bellow out, 'Look! It's the fucking firemen here to piss on our lovely fire!'

Her equally squalid-looking comrades chortle out a drunken cry of 'Piss off . . . effing leave us alone,' waving their cans of strong drink and bottles of Red Biddy. Experience has shown us how lethal such a liquid can be, with victims having set themselves ablaze by rolling too close to the flames under its influence.

In any event, their bonfire this evening is too big and is sending up a load of sparks and embers that only need to find their way through the open window of one of the adjacent commercial premises to start an inferno. We do, however, appreciate that it is a freezing cold night and, for these people, their fire is the only warmth they have. If we douse it completely, they will light another one a few yards away. In our experience, the most prudent compromise is to reduce its size.

As our Station Officer, Jim Cronine, a burly ex-Royal Navy sailor, approaches the group and the fire, one of the men, his senses befuddled by the potent drink he has consumed, grabs a flaming brand and waves it at him, shouting in a broad Irish brogue, 'I'll effing brand you if you touch this fire!'

'Now, now, let's not do anything we might regret,' booms the guvnor, striding purposefully up to the drunken Irishman. 'We are not going to put the fire out, just reduce its size, Paddy. OK? Understood?' he says, his 6 ft 1 in. height and 17 stone build towering over the down-and-out.

'Well, make sure that's all yer do,' he snarls back, a spray of alcohol-tainted saliva accompanying his words.

We soon reduce the fire to a level we can live with, giving it a few bursts from the hose reel and breaking it up a bit with a ceiling hook. Within minutes, we have wound the hose reel onto its drum and a quarter of an hour later are able to finish off our boiled beef and carrots, which, having spent an extra half-hour in the oven, is by now more boiled than we would have liked.

It's just after midnight and there are six of us sitting at the Formica-topped table in the mess room. The room, about 18 by 9 feet, is crying out for refurbishment; the cream paint is peeling on the walls of this particular Victorian-era fire station. There is an old radiogram in one corner and next to it, a battered wooden bookcase with a sliding glass door, containing a range of reading material, some of it fire brigade-orientated, but there are also various paperbacks and fiction, plus a pile of old Playboy magazines, the latter's pages dog-eared after three shifts of firemen, who have scanned the photographs with voraciousness or nonchalance, depending on their age and attitude.

It is a habit for men to congregate round the mess table, sipping mug after mug of tea or coffee. I always preferred to be alert and conscious should the bells come in, and there were many just like me who chose not to take some fitful rest (which we were all permitted to take) on our army-type iron-framed beds in the shabby dormitory.

'I reckon it's better to stay up at night, rather than to jump up from a sleep when we get a shout,' says Jim Laine, a thickset veteran of 22 years. 'I read somewhere that the average fireman only lives about five years after retiring at fifty-five and I wouldn't mind betting that it's the sudden action from rest to firefighting that starts the rot,' he rattles on.

'For God's sake, Jim, leave it out, you morbid bastard. That means I've only got six years left if I retire next year,' responds Bert Braxton, at 54 one of the oldest men in the division.

'Yeah, but you've probably spent half your career sitting in here after midnight, so the old ticker won't have had the strain put on it like what happens when you're suddenly woken by the bells. Christ, take a look-see next time we turn out at night at those blokes who always like to kip – they look like death warmed up,' Jim continues.

Most of this 'club' of nocturnal tea and coffee guzzlers are still around the mess room nearly three hours later. The air is now a fug of dry heat from the radiators, fuelled by the coke-fired 'appy' – the affectionate name given to the heating furnace apparatus ensconced in the basement – and from the blue haze of cigarette and pipe tobacco smoke.

Few at the time would have been able to envisage a future where most organisations, including the Fire Service, are so reformed in their culture of health and safety that smokers are in the minority. Bert Braxton and Jim Laine had, in the raw days before the advent of breathing apparatus for every crew member, probably inhaled a far greater volume of lung-damaging fumes and smoke during hundreds of fires than a whole room full of smokers could ever match. As a consequence, they wouldn't have given even a thought to a colleague exhaling cigarette or pipe smoke in their presence. Indeed, cigarette and pipe smoke was considered a natural part of that cosy warm fug of the fire station rest area, by smokers and non-smokers alike.

Three of the six men remaining have whiled away the last couple of hours playing cards. Another has stretched out in the battered old armchair with a novel, whilst the two veterans of the station have indulged in their nightly reminiscences of fires and characters, looking back over a long period spent within the London in the East.

Only about a half-mile away from where these six night owls are sitting, dry and warm, and where the remainder of the watch are dozing fitfully in their iron beds on what has been an unusually quiet tour of duty, another six people – four men and two women – are huddled and slumped in the rat-infested remnants of a once-proud house.

Brick Lane, London E1, is a slightly curving thoroughfare of about half a mile in length. At its northern boundary it makes a junction with Bethnal Green Road. Travelling in an easterly direction are such streets as Cheshire, Buxton and Old Montague, which themselves run into Vallance Road, the one-time home of three men, two of whose names, Ronnie and Reggie Kray, have become synonymous with the East End of the '60s and '70s.

At its southern end it becomes Osborn Street for a short stretch before meeting the major highway of the Whitechapel Road. Within a little over half a mile on the left is the Blind Beggar public house, reminding us again of the infamous Krays, for it was in this drinking place that Ronnie Kray killed an old enemy, George Cornell, by means of a close-range bullet to the head. Some 80 or so years earlier, in another infamous period, which again helped to create the folklore and history of this district, a certain slaughterer of prostitutes gained renown as Jack the Ripper.

Between the 1970s and the 1990s in particular, though it is still true in some places today, save for the replacement of gaslights with electric ones, many streets and alleyways in the area could have passed for the grey-stone, soot-blackened redbrick buildings and cobbled, stone-setted streets of that period. Brick Lane and its surrounds had witnessed much of the work of the London Fire Brigade over its long history, during which a succession of immigrants, including the Huguenot, Irish, Jewish and latterly Asian communities, had occupied the houses, tenements and factories in successive waves, most striving to make a living, and in some cases their fortunes, through a variety of trades but especially centring on the garment and furniture industries.

If there is such a thing as spirits from the past, then, in fire brigade terms, the atmosphere there is dense with them: the clatter of horses' hooves as they haul the steamer fire pumps; the cries of the firemen drivers urging their horses on to a blaze. Then, in later years, the haunting clamour of bells and sirens: the orange, then later blue flashing beacons: the drone of powerful diesel engines pumping water onto fires in a multitude of factories, workshops, villas, houses and tenements – fires whose thick smoke will have drifted over the rooftops of this cosmopolitan inner city district, dispersing its distinctive, sharp odour and draping its hot, twirling cloak over the homes of the hundreds lying asleep. Asleep and unaware of the drama taking place in the shadows of the East End night and of how that darkness is now being rapidly illuminated by the lurid glow of leaping, searing flames.

The bone-chilling coldness of this particular night, exacerbated by a biting wind from the north-east, must have forced the inebriated inhabitants of Itchy Park whose fire we had dampened down earlier to seek out some shelter from the elements. They headed for one of the derelict buildings in the minor grey backwaters branching off Brick Lane. It's no more than a quarter of a mile to Brick Lane from the site of their earlier bonfire, the embers of which were now slowly turning black as the fire cooled from its abandonment. Just off Brick Lane, they arrived at a partly boarded-up house whose crumbling walls and holed floors had provided sanctuary on several occasions in the past.

The subsequent fire investigation report and interviews with the lucky vagrants who survived the blaze painted a colourful picture of the events that preceded the call-out to the derelict building later that evening.

The men and women of this group all said that, as they were heading inside, the small grizzled Irishman who had threatened to 'brand' our Station Officer some four hours earlier shouted out that he was going back to the soup wagon outside Christ Church. With a snarl and a curse to the cold, the whole group of six, their old dirt-stained coats flapping in the rising wind, and their worn-out and burst shoes dragging over the hard London pavements, reeled off, clutching their bottles and paper bags of wine and spirits. They eventually reached the mobile 'kitchen' and huddled around the stark white light of its illuminated serving hatch like moths around a flame, the bright light standing out like a beacon amidst the dark alleyways and courtyards of the famous market opposite.

Hot onion soup and ragged slices of bread were on the menu that night, and by 1.30 a.m. the group had filled their stomachs and made it safely back to the derelict house. A box of matches was then pulled out of the greasy inner pocket of a ragged gabardine raincoat. A small pile of paper and cardboard was pushed under a pile of broken bits from a picture rail and was lit.

It was only a small fire, but, since the fireplaces were all boarded over, it had to be positioned against a wall of a rear room. This position also disguised their presence from patrolling police officers who might notice the shadows cast by the dancing yellow flames; the smoke, however, which soon found its way out of the few upstairs window openings, could still be detected if the now gusting wind were blowing in the right direction.

As the warming and soporific waves of heat enveloped them, their fingers, numbed by the low temperature, began to thaw. Soon bottles were at lips and over the next few hours the six easily drifted into a slumber, their bodies slowly sinking onto the broken timber floor covered in crumbling plaster, broken laths and skirting boards.

Before finally succumbing to the mind-numbing effects of half a litre of Red Biddy, one of the group added some of the old skirting to the fire, placing it in criss-cross fashion over the hungry flames, whose appetite, whetted by the old melting layers of paint, caused them to stretch their hot tongues up to the timber.

With each passing second as the six down-and-outs snored and grunted from their chosen positions around the ground and top floors of the three-storey house, the flames lengthened and grew hotter as they consumed their fuel.

London's streets are, of course, the workplace of the police officer and it was one of them who first caught the sharp tang of burning in his nostrils just after half-past three. He knew the direction from which the biting wind was blowing and began to walk towards the smell of smoke – slowly at first, then faster as the pungency strengthened. As he approached the street's junction with Brick Lane, his heart leapt. Swirling wreaths of smoke were racing past the pale yellow of the street lamps and the smell was now so strong that he could almost bite on it.

Jim Laine and Bert Braxton have almost covered 51 years of their combined service during their nostalgic reminiscing when the bells come clamouring in. It matters little that experience has taught them that many calls turn out to be not quite the dramatic incidents they envisage because you never know the real score until you arrive. Consequently, the heart begins to pound with a surging, stomach-tingling mix of excitement and apprehension. Such emotions are at their greatest when the call-out message contains the prefix 'Fire' or 'Smoke Issuing', and it arrives in the early hours.

It is dead of night now and the teleprinter punches out '*Fire in derelict house – Brick Lane E1. Time of call: 0355*'. No more than three minutes has elapsed when the officer in charge of each appliance sends a 'book in attendance' by radio. The thick grey smoke of the fire is illuminated by their headlamps and reflected back like a dense fog, in spite of the strong winds blowing spasmodically.

When a sudden gust clears the smoke momentarily, in front and on our left a heavy body of fire is showing in the second and third storeys of a typical early Victorian house. The red and yellow glow at the ridge of the house indicates that the fire is 'through the roof' – not unusual in houses fallen into dereliction where rotten timbers, plus the breaking up of rafters and purlins by down-and-outs to build fires, have opened part of the roof to the elements. This is a good thing, in the sense that the fire and smoke can vent: conditions of visibility and breathability improve for crews working inside and vital air can reach anyone trapped within the building.

As this is an area of commercial premises intermingled with tenements and houses, with a fair proportion being in various stages of dereliction and disrepair, there is not the same sense of urgency when searching the building. Although we know from experience that in such parts of the capital down-and-outs might be sleeping inside, our automatic search routines, which we deploy immediately at conventional residential fires during sleeping hours, can be easily pushed into second place at fires in derelicts.

Within this high-risk tinderbox of an area, though, suppressing fires is crucial if spread to the densely packed commercial premises all around is not to occur. The body of fire is such that the two-and-three-quarter-inch-diameter hose is the preferred choice for a quick knockdown. The gallonage of water delivered per minute via a suitable-sized nozzle means that the maximum absorption of heat, combined with the weight of water striking the flames, is very effective.

I fling open a side locker, grab a length of the large hose and throw it along the road, where it unrolls with a slap. Just as I am connecting the coupling to the delivery on the pump at the rear, a terrible scream cuts through the air from above.

I look up. There, at the very centre of the window frame on the top floor, the form of a man, his clothes alight from head to toe, is silhouetted by the fire in the room behind.

'Send a priority "*Make Pumps Four – Persons Reported*",' the guvnor shouts out. 'Slip and pitch, plus a covering jet,' he continues breathlessly.

At 54 years, and with only a year left before he will be forced on grounds of age and rank to take compulsory retirement, Bert Braxton could choose without feeling ashamed to use all of his 'ringcraft' to ensure that at most fires the heart-pounding stresses and lung-and-leg-bursting efforts of rescue are left to the younger element. But he is a fireman's fireman. An individual who is just as motivated here, at this ungodly hour, when confronted by the sight of a burning man 40 feet atop a virtual funeral pyre in the heart of London's East End as he was as a young fireman all those years earlier.

Back then, he had first felt the surge of adrenalin – that hard to describe 'buzz' – after bringing three small Jewish children down from the upper floors of a burning tenement no more than a few streets from where we are now. In that rescue, Bert had ascended and descended the wheeled escape three times to rescue those screaming, sobbing children. The ladder was almost enveloped in the rolling, rising flames issuing from the lower floors. Even though a young man at the time, with the natural resilience of his age, the effort and energy of those rescues in the deepest hours of the night had all but laid him low. The Bert Braxtons of the Brigade never sought glory or reward – fire and rescue was their noble calling; human life was precious – so it was good, therefore, that the Honours and Awards Board of those far-off days had seen fit to commend his heroic actions as above and beyond the call of duty. The framed certificate on the watch-room wall spelled out his bravery that night, a fitting reminder in later years of the spirit and courage displayed.

Although time is always of the essence in firefighting in general and in rescues in particular, Bert knows that a man can move too fast when smoke and fire are about. He knows this from experiences gained at hundreds of emergency attendances over nearly 30 years, all of it spent on inner-city stations. Such experiences have taught him and his colleagues how to conserve that energy and strength so vital to the saving of life.

His ascent of the now pitched ladder is a master class in how to move briskly but not recklessly, conserving his breath and reducing his need to pull in huge amounts of air and smoke, as would have been the case if he had run too hard up the huge 50-foot rescue ladder.

Bert's nocturnal reminiscing partner is now partnering him

again. Jim Laine, another veteran, is providing the covering jet and his first task is to shout up to the screaming and burning male at the window. 'Stay put, mate – we'll have you in a second,' after which he allows the powerful jet to crash into the brickwork just above the head of the unfortunate man, the icy water deluging the burning clothes.

Bert – calm, controlled, but with his 54-year-old heart thumping like a pile driver against his breast bone – is now at the head of the escape. Moving with an agility that would shame men 20 years his junior, he clambers into the burning top-floor room.

The water from the jet has virtually extinguished the burning garments, but the man, still in a drunken haze, and traumatised by his ordeal, is reeling from side to side. Bert can see that there is no way out of the room. The flames, heat and smoke are too much, so without hesitation he bends his near six-foot frame over. In a flash, the small Irishman, the very same who had threatened to 'brand' our guvnor the previous evening, is slung across his shoulders like an uncomfortable, filthy stole.

With burning skin from the man's hands and legs filling his nostrils with a sickly, hot odour, Bert remounts the ladder and executes the classic fireman's lift carry-down, to where willing hands relieve him of his badly burned human burden.

'If there's one in there, then there are probably others,' shouts the guvnor to Sub Officer 'Ziggy' Apted. 'Get a jet in via the front and get the BA men rigged.'

After connecting the large hose to the pump delivery, I take a standpipe and a hydrant key and bar, and locate the nearest hydrant. From this, we lay out another line of hose to provide an unlimited supply to the pump. By this time, the escape has been pitched to rescue the blazing man and we are ready when the Station Officer orders the large jet into the house via the front door. Although locked, it soon succumbs to the weight and blows of the 14 lb sledgehammer being wielded by former heavyweight boxer Jim McBreen. As Ted Smart and I hump the now charged hose line up the small flight of stone steps that lead to the front door, I can see the BA men rushing to rig. To save time, always of lifesaving essence when persons are reported, two men have entered without the protection of BA and these two are quickly inside, crawling around the smoke-filled house searching for any other persons. As the fire has vented through the roof, the smoke and heat are a little more bearable than is usually the case.

As we haul the heavy hose into the ground floor, it becomes clear that the fire has apparently started on the ground level but has bypassed it to create fire on the top two storeys. We soon have the charged line up to the second floor, where there is a heavy body of fire in two rooms and a hallway to the front of the house. As the powerful stream of water hits the ceiling and walls in its ever-moving sweeps, we are drenched with a deluge that is already warm because of the heat absorbed in its cooling and knocking-down effect.

In later years, when looking back at so many working fires, like this, that we responded to in those times, I never fail to be surprised at the amount and minutiae of detail I am able to recall of the premises involved.

I can recall the broken wall and the ceiling plaster with the broken laths; the patterns on the linoleum, stained and worn with rain from leaking windows and roof but which had once been shiny and new; the remnants of furniture from another era, now standing sad and abandoned in a corner of a smoke-laden room. Above all, I can recall the past spirit of lives moved on: the bedrooms of love, life and death; the dining rooms or kitchens; the food eaten before children danced off to school and parents to work, many in the clothing and furniture trades that then made up such a large part of this cosmopolitan inner city's lifeblood. Spirits of the past. Memories, which not even this fierce fire's flame and heat, could ever erase.

Our two colleagues, Johnny Depthard and Mitch Rolinger, who had responded to the Station Officer's urgent request, have done a search of the ground floor made hazardous by its missing floorboards and general state of dereliction. Their bare hands had recoiled from the hot embers and ashes of the fire set by the squatters some four hours earlier, which had grown out of control. The heat and smoke had risen to the ceiling and, from there, had raced up the hallway and the stairs, its superheated gases raising the temperature of the dry timbers, old paint and wallpaper of the upper floors to the stage where they burst into searing flames.

'I've got somebody here!' coughs Johnny Depthard, as his hands fall onto the unconscious shape of a body. 'Grab the legs, Mitch, and drag it out back to the street, quick.'

Retracing their route in, they manhandle the dead weight across the broken and litter-strewn floors to the front door and down the unforgiving stone steps to the pavement, just as the BA crews are going in to take over the search-and-rescue effort.

The body belongs to a man. He is dressed, if that is the right word, in a greasy, once sand-coloured gabardine raincoat. It's now almost brown with the polished shine of too much wear – for too many days and long nights had that raincoat, fastened round the waist with a section of brown cord, acted as the only piece of material between his malnourished frame and the cold rough floor of a derelict building. His trousers are corduroy. Once green, now almost black, with frayed bottoms and shiny with wear like the coat. He must have taken off his shoes before lying down to sleep and on his feet are the black remnants of socks through whose stinking smell poke his big toes, the nails of which are black, long and curled back into the skin.

A quick check of the carotid pulse in the neck detects a still beating heart and so the London Ambulance Service takes over the next stage in the rescue and revival process.

We have soon knocked down the fire on the second floor and, in spite of the noise made by the powerful water jet and crashing plaster and hissing steam, we can hear the welcome sound of braying air horns, which herald the imminent arrival of the reinforcing appliance.

Within minutes, we have negotiated the staircase. In a short time, our 60-gallons-a-minute of water has put paid to the inferno in the room from which veteran Bert Braxton carried out his carry-down rescue with such aplomb only a few minutes back.

The BA men locate four other occupants on the ground floor, however because the fire has bypassed them, its fury being diluted by the open roof, allowing heat and smoke to ventilate, their unconscious state is more a result of the alcohol consumed earlier than from the malevolent smoke surrounding them. It is most likely, though, that had the policeman not detected the pungent smell of burning and called us, they would have ended up on a mortuary slab rather than the pavement of Brick Lane to which good fortune and the well-drilled search-and-rescue efforts of the London Fire Brigade have mercifully delivered them on this bleak winter morning.

'Down-and-outs' brought 'out and down', as so many were during those busy years when fires surrounded the beating heart of the East End.

It is not always a good thing to reflect on the outcomes in life for those who have been caught up by the hazards of fire or other calamities. After all, we are not there to be social workers and if a man's social conscience is too strong, or if he has such a strong conscience but cannot suppress it, then he might be in the wrong occupation. This having been said, sometimes we would wonder amongst ourselves, as we sat around the mess room gulping sweet hot tea after a nasty or difficult job, where such people as these down-and-outs ended up. After attending more than my share of fires involving such dispossessed persons, I came to the conclusion that, with a few rare exceptions, they continued to live rough. There are a multitude of reasons that put a man or woman over the edge: thwarted in love, a broken marriage, alcoholism and violence, bereavement, drug addiction . . .

Most of those we rescued from what would have been an otherwise multiple-fatality fire would have been checked over at the hospital, cleaned up and discharged, and maybe pointed in the direction of a local hostel. The man who was rescued and was quite badly burned would of course have been admitted and treated like any other person, but if my experiences of vagrants have taught me anything, it is this: if he healed without complications, he would come out a lot cleaner in body than when he went in. Although it is highly probable that within a short time you would find him back again in those derelict buildings and back alleys, living rough and imposing his own death sentence through his lifestyle – I once read that those who imbibe methylated spirits can expect to be dead in months if they continue.

What such people have to be thankful for are those people who work for the Salvation Army or other charitable volunteers who go out of their way and give so much of their time to ensure that those dispossessed within our cities have some warm food and drink and, if lucky enough, a warm, dry bed rather than a cardboard tent or the hard floor of some derelict rat-infested building.

Those who give such aid are the real heroes, in my book. We get a salary, they do it for free because they are blessed with a high level of humanity and understanding.

Chapter 18

Perilous Petroleum

A Cruel Price

It's well past two in the morning on the first of our two fifteen-hour night duties. I am sipping strong coffee, the standard beverage for those who, like myself, find sleep hard to come by even with the inducement of a decent mattress and pillow and the couple of rough grey blankets the brigade provides.

There are ten on duty tonight. The 'guv', Biff Sands, kips in a separate ground-floor office next door to Sub Officer Jack Hobbes and Leading Fireman Dick Friedland. Of the other seven, three – Ricky Tewin, Lofty Morphard and I – are spinning yarns about some of the most memorable incidents attended at the different central London stations on which we have served for over twenty years between us, while fellow Yorkie Barry Priestley, Toby Smollett, Jerry Pashleigh and Paddy Mulligan are in their pits, all being the sort who could sleep on a clothes line. Two of them, with big families, also moonlight in between shifts and on the rota days to boost their incomes, but there is a rumour that this might soon be clamped down on, owing to the health-and-safety concerns of the higher brass.

'What's your take on the rumour going round that the brigade will be included in the proposed national health and safety at work regulations?' Ricky Tewin asks no one in particular.

'My take,' replies the 6 ft 4 in. Lofty, 'is that it's a load of cobblers. How on earth can the Fire Service, whose recruits enrol knowing that they will be expected to be placed in serious danger by protecting life and property, be strangled by such regulations? It's bleeding—'

His reply is cut short as the teleprinter bell strikes up. At this early hour, its actuation can be caused by only two things: an incoming emergency call (the delay in the call bells due to the address being sought from a panicking 999 caller) or an ordering for the Pump on a non-blue-lights journey to provide a relief crew at an earlier incident somewhere in the capital.

Seconds later the darkened appliance room is suddenly bathed with automatic lights as the wall bells begin their loud and urgent tolling. It is a shout and we all spring up from our chairs like a jacks-in-the-box, plunging ourselves down the cold steel of the pole. It is 3.30 a.m. now, not a good time for fires, as most residents are unaware of a growing blaze that might be placing them in the long sleep of death.

The shout is to a part of the East End that, following my recent visit to New York City Fire Department, I liken to Manhattan's Chinatown on the Lower East Side. True, we haven't the external iron fire escapes that festoon the tenements there, but a lot of the buildings are of a similar appearance and the risk to life from fire is not much less.

The two appliances squeal to a halt in front of a seedy redbrick tenement in one of the meanest streets of this down-at-heel district. The building is obscured by a thick pall of brown, acrid, nose-searing smoke but before we can really register this a horrific sight confronts us.

There is a man or woman – it's too difficult to be sure in the thick fog – reeling around with their clothes alight from head to foot. The person is screaming like a banshee as the flames devour his or her skin. Clearly, there is serious fire about – we know that temperatures of 300°F can be reached in these fires, causing second-degree burns on exposed skin in as little as 15 seconds. No wonder this sorry individual is squealing like a stuck pig.

Lofty's long legs catapult him to the flaming person in a lightning flash. He rips off his axe belt, tears off his fire tunic, pulls the victim down onto the cold street using the tunic as protection and smothers the angry, hungry tongues of yellow flame – but not before this poor person's windpipe has been seared by gasping in the blistering heat.

'Get a jet to work now!!!' roars Biff as he sizes up the tenement from which rolling clouds of brown and yellow-flecked smoke are billowing at great speed – worrying signs of a really hot and nasty inferno.

I dash past a now gathering motley bunch of spectators: neighbours wakened by the blood-curdling screams and the noise of our arrival probably. Some have a blanket over their night attire, but a few men stand clad only in trousers, string vests and slippers. I notice a gaggle of blousy buxom blondes in half-fastened towelling gowns, all of them transfixed by the high drama. It might be the meanest of mean streets, but even this dodgy district doesn't experience fires like this every day.

I fling open a side locker, grab a length of small diameter hose and sling it along the road, where it unrolls and slaps onto the damp cobbles under the smoke-obscured light of a street lamp. I hear the satisfying click as I plug in the hand-controlled branch pipe, drag the uncharged hose towards the front entrance to the tenement and feel the satisfying surge of water as the flat snake engorges and tightens into a hard tube as driver Toby cracks open the delivery valve of the powerful pump.

Jerry and Paddy are rigging in BA as Ricky and I go low underneath the super-heated band of rolling smoke and knockout-punch fumes. We have advanced no more than 15 feet into the blazing hallway when from out of the thick, hot veil comes another awful scream. It is a second person, fully enveloped in flame, staggering towards us.

I quickly activate the powerful spray and its cone douses the poor sod, who collapses with a long moan, as hair and clothes and terribly burned, hanging skin are cooled. Biff has called for another Pump and his 'persons reported' message will see at least one ambulance on a mercy dash.

Two men get burns sheets onto the stricken pair, one of whom we discover is a male, the other a female.

'Effin hell,' mouths Ricky, as we jointly advance, his arms hauling the line as we crawl into this baking heat along the tiled floor of a longish passageway towards the menacing maw of flame some 15 feet further along. 'What the fuck have we got here?' he coughs, the acrid smoke hitting him hard in the nose and throat.

'Looks like somebody has torched the place,' I manage, feeling my head becoming light as a result of the heat and smoke.

We reach the burning room, which is pushing out rolling red flames at ceiling level from the door of an apartment to our left. These life-extinguishing flames are being attracted towards the outside air like iron filings to a powerful magnet. I sense a cold shudder even in this furnace, having seen this scenario before: this flame-tinged brown and yellow smoke indicates perilous conditions, a real firetrap on the point of turning into a fireball that could see off whoever is in its deadly path.

Instinctively, I put the jet into its fullest bore and turn into the door on our left, directing the surging torrent up into the rolling flames of the ceiling, hearing the crash as it dislodges fire-weakened plaster and lath. I wince as a wave of super-heated steam envelops us, the jet doing its job of suppressing and cooling the cruel tongues of flame.

Behind us the thud of leather soles and the muffled grunts of the BA men indicate that they are with us and preparing to search the higher floors, as we concentrate on stilling the raging beast that has caused such terrible havoc.

As I touch the hot floor of the room, my left hand lands on a small mound of what feels like hot, soft cardboard. Most of the main body of fire has been suppressed now, so, curious at the sensation, I shut off the branch and place it down, keeping crouched in what is still a baking temperature – God only knows how hot it must have been before we arrived. Sweat is oozing from every pore and my eyes are stinging, like when a hot chilli pepper seed from one of Carmel's curries accidentally gets into my eye.

I turn on the weak yellow light of the hand lamp that is clipped to my axe belt, point it downwards and the supper of seven hours earlier comes into my throat at what I see. It is the terribly burned body of a child. Going by its size, the victim is probably no more than two or three years old, and I gag at the sight of the awfully seared skin of the tiny face and the horribly blistered little body. Into my mind come the words of a poignant poem penned by one of the brigade's senior brass and recently published in the brigade's monthly magazine: 'Little children who suffer, in the sense of early hours. God give us strength to save them. Stop the Devil picking flowers.'

In due course, we return to the refreshing coolness of the street, but with a heavy heart at the awful sight witnessed. No matter how many times I was involved with death, especially that of children cut down before they had even blossomed, I never got used to it. But we are inner-city firemen and this is what it is like sometimes; we have to cope and reassure ourselves at such times with the fact that we save more than we lose.

We are a solemn bunch around the mess table a couple of hours later. Solemn but very angry when we hear that the police think the fire had been a wicked and cold-hearted act of revenge for a financial debt unpaid – the cruel price being the lives of a tiny child and its loving parents.

Ladies of the Night

Any well-researched history text on the inner East End will reveal the more sordid side of life in and around the London docklands and its neighbouring districts. All the world has probably heard of Jack the Ripper and the prostitutes he slaughtered in the latter years of the nineteenth century. Seaports of the world's cities with a constant influx of transient mariners, starved of the pleasures of the flesh for weeks on end, will always participate in the oldest trade in the world. When some thought is given to the fact that many men who use the services of prostitutes are complete strangers in many cases, the risk to their personal safety makes the trade more hazardous in some ways than being an inner city fireman.

This is especially the case for those who work the streets, or those call girls who work from home without the presence of a

Madam as a security back-up. London's East End was no exception to this rule. In the immediate post-war years, a lot of the buildings of the Blitz-ravaged districts, especially those around Shadwell, Ratcliffe Highway and Cable Street, were opened as cafés that were open virtually 24 hours. 'Café' was in most cases a front for a brothel and the dock traffic, along with those men seeking a sexual outlet from elsewhere, provided a never-ending trade.

Even though in the early 1970s there was not the same proliferation of the business as there had been in the previous decades, there were districts and streets that were still notorious. It is to one of these that we have been called to a report of '*Fire – persons believed involved*'.

It is 4.30 a.m. and as we near the address there are several ladies of the night patrolling their patches, teetering on their high heels, with low-cut tops exposing their cleavages in spite of the chill. All appear to be oblivious to the wreaths of thick, drifting smoke swirling around the pale yellow of the street lights.

Both appliances halt in a mean street of dilapidated terraced houses adjacent to a steel girder bridge over which a suburban railway line runs and on which the words 'Asians Out' have been roughly painted.

As we jump down to the street, a piercing shout for help hits our ears. It is coming from the end house in the terrace, from which brown wreaths of smoke are percolating through the front entrance and from the roof area. The searchlight is quickly pulled from its mounting atop the Pump, its white shaft of light soon illuminating the roof. I grab the nozzle on the side-mounted hose-reel tubing and yank hard, pulling off a good 50 feet, ready to take it in as a first attack on whatever is burning.

Toby Smollett is driving the Pump and he soon locates the yellow-and-black hydrant-indicator plate. He takes a standpipe from a side locker and within seconds has connected into the town's mains to give us that precious, often lifesaving water.

'Send a "*Persons Reported*", Toby,' shouts Biff Sands, as the searchlight picks out the source of the screaming. It is a woman half out of a skylight on the slated slope of the roof, 35 feet above. Smoke is issuing from the roof opening and she has got one long, bare white leg out of the skylight and onto the slope, with the other still inside. While the BA wearers are rigging, I take the hose-reel nozzle and head for the front door, as Ricky Tewin feeds out the red rubber tubing from the festoons I had previously pulled off.

The front door is of a heavy wood construction and is locked, but rugby playing Martin Bauer makes light work of forcing an entry, wielding the 14 lb sledgehammer in a fashion that would make any Irish navvy proud.

The escape is pitched to the gutter level in case we cannot get to the trapped woman via the internal staircase or before the fire lights up and traps all inside. There is heavy, hot smoke in the hallway but only a small amount of flame coming from the floor and lower part of the staircase. I crawl in very low below the smoke and squeeze the nozzle's trigger, feeling the satisfying recoil as the powerful small jet bursts forth. The flames disappear rapidly and the smoke banks down – I take in an unwanted lungful that has me coughing violently.

Dick Friedland and Jerry Pashleigh are fully rigged in BA now and they brush past me onto the dog-legged stair and up to the top floor, where the victim was seen. Within a couple of minutes, the heavy thump of boots on the staircase indicates their return. I have knocked down the fire now and windows are being opened to ventilate, as Jerry appears through what is now not much more than a haze. Over his right shoulder is a voluptuous female form. She is wearing black fishnet stockings, a red baby-doll negligee and a black leather thong, which accentuates her plump, goose-pimpled buttocks. She is semi-conscious now, either from the smoke or from fear, drink or drugs – or perhaps a combination of them all. Her peroxide blonde hair is hanging down Jerry's back and draping over the black-and-white cylinder of his BA set, which sits in its harness in the small of his back.

Biff Sands' white-helmeted form appears. 'Have you got it, mate?' he enquires. I cough out an affirmative, wiping the snot from my nose onto which my watering eyes have run.

It looks like the seat of the fire has been just behind the front door and has spread via the combustible floor and stair carpeting.

I sense that someone might have poured some accelerant through the letterbox and dropped a lighted rag or something to ignite the vapour. They probably didn't use sufficient accelerant or there would have been the same sort of inferno as seen on previous arson incidents.

I catch the guvnor on the top floor and let him know my suspicions; he has formed the same conclusion.

'We've had quite a few jobs in this neck of the woods. There are some right friggin lowlifes use this street, but you should have seen it 20 years back, especially in and around Cable Street,' he rattles on, as we enter a room just off the top-floor landing.

It is a large red-wallpapered bedroom of about 20 by 20 feet. Bang in the centre is a huge cast-iron black-painted bed, atop which is a red-and-gold bedcover with matching tasselled pillows. There are a pair of handcuffs fastened over the bed rails and a similar pair of ankle cuffs at the bed's foot. A pair of wall-mounted imitation candles with red bulbs illuminate the bizarre scene.

Attached to the ceiling is a huge mirror, and there are various pairs of leather, stiletto-heel thigh boots lying around the floor, plus two black nine-tailed leather horsewhips in the centre of the bed.

On a red dressing table opposite the bed, there is a small glass bottle with 'Amyl Nitrate' printed on a label.

'Bleedin' hell, mate. I've been to a few fires in knocking shops around 'ere, but I ain't never seen this set-up. It must be one of them tarts whose cards you see plastered inside the telephone kiosks, "Miss Correction-Dominatrix-Discipline and Bondage" and all that,' Biff exclaims, taking off his helmet and scratching his head.

'What's with the amyl nitrate, guv?' I enquire.

'You're the young geezer, so I thought you could tell me, my old mate,' he replies, thinking for a minute, then saying with a grin, 'Amyl nitrate . . . I've read somewhere that if you sniff it when you're at the old sex malarkey, it increases the sexual sensation, I think.' There is a pause before he continues, 'But I might be wrong, so don't bleeding quote me, yeah?' I suspect choosing his words so that he can't be incriminated and endlessly ribbed by the rest of the station, the suggestions being that he only knows because he's tried it!

Back down on the street, the blonde prostitute is in the back of an ambulance sent by Fire Control on receipt of the 'persons reported' message. Police officers are also inside the ambulance, the doors of which are wide open. Apart from a bit of smoke inhalation, it appears that our 'Miss Correction' is otherwise unharmed.

It transpires that a punter much earlier in the evening, in spite of all the tools of sexual titillation present, not to mention the Amyl Nitrate, had not achieved any gratification and had demanded his money back. He had been refused and left vowing that she would be made to pay.

She had gone to bed in the small attic room she used as a bedsit, only to be awoken by a noise, probably the heavy iron letterbox flap being opened. Within minutes, the house had filled with smoke, her only escape being through the roof skylight.

Police forensics attended the scene and their accelerant detection probe picked up strong petroleum vapours under the letterbox, so it appears that petrol had been poured through and ignited, and the matter was now one for the boys in blue.

Remembering other fatal consequences of earlier arson attacks, we all agreed that 'Miss Correction' was one lucky bunny. Another half-can of petrol and it is highly likely that in spite of our rapid response and highly honed rescue and firefighting skills, it would have been a blackened corpse we would have been dealing with. Such are the hazards in the bizarre world of sex for sale. That night the punter's luck was out, whilst hers, fortunately, was in.

Arson by Persons Unknown

It's Friday, the second of our 6 p.m. to 9 a.m. night shifts. The previous night and the previous two day duties hadn't seen a wheel turn in anger. The lull before the storm?

I am slumped on the iron-framed dormitory bed, staring up at the ornate plastered ceiling. It is just before two when I feel waves of sleep enveloping me. My senses are usually too keyed up with the anticipatory tension of knowing that the huge, red call bell on the dormitory wall could sound at any second to sleep soundly, so any sleep I do get is the sort that leaves me feeling tired next morning.

In my late teens in the north just before I became a fireman, I can recall looking across to darkened fire stations in the inner-central suburbs of such large cities as Leeds and Manchester as we drove home in the early hours after a night on the town. As I looked, I knew that if things went my way I would soon be part of this vital lifesaving occupation and working within similar buildings, awaiting the emergency call in those wee small hours. Back then, I was always struck by the contrast that existed in those early hours: there being no signs of life in the barely lit station building, with appliances just visible through the glazed red doors, then the transformation when a call came in. The darkened premises instantly became a brightly illuminated scene; where a second earlier things had been so still, the next it was a frenzied hive of urgent activity, with men dashing across the vehicle bays before clambering aboard.

Blue beacons would spin, doors would crash open and Pumps would roar out, leaving a pall of diesel smoke, the call bells often still ringing, so swift had been their urgent departure. As they raced away, I was often left curious as to what they might soon be facing and if they would return safely.

It is the dead of night now. Other than the dim yellow light within the exterior 'running call' telephone box, the station is pitch-dark. Perhaps some young man with the same keen fire service ambitions as I held is at this moment passing by and looking across to the darkened station as I once did.

Then suddenly the bells are tolling, and the dormitory and appliance room are bathed in light. I glance at my wristwatch as I throw back the grey top blanket. It is a quarter to four.

Both red and green indicator lamps are lit and, with a pull on the rope, the pair of huge red doors slam open and the night air rushes in as we leap onboard. I am riding the Pump Escape.

We had done several drills with the 50-foot escape the previous evening, honing our skills of slipping the one-ton wheeled ladder from the Pump, manhandling it into line with the drill tower's fourth-floor window opening and strenuously winding the extensions to the required height before dropping the white-painted head of the ladder onto the wooden sill.

Our Station Officer, Biff Sands, is a firm believer in the value of repetitive drill evolutions with the escape. For him, this stems from his army days and the repetitive practices of assembling his Bren gun during the Korean conflict. He knows that it is only by doing drills again and again that actions become automatic and spontaneous when the chips are down.

No more is this trained spontaneity as necessary as at four in the morning when some of the crew are still on autopilot after their sudden rousing. The conditioned response brought about by regular practice can be the difference between someone being saved or lost in a serious, rapidly developing blaze with persons trapped on upper floors.

It is no more than five minutes to the address of the call, a report of '*Fire*' in a long street of four-storey Victorian terraced houses so typical of the inner London scene. The crew cab window is half open, both to better ensure we are fully awake and to pick up any scent of that peculiar smell that accompanies a burning building. That familiar acrid odour is definitely in the air and it is very strong. Even if the inrushing cold air doesn't do its job of fully rousing men, the pungent smoke usually does – no matter how many times its smell has entered the nostrils – and hearts beat faster in anticipation of a working job.

As we race into the street, we are left in no doubt that this is a working job and, in fireman speak, it is a 'right goer'. There is a terrific amount of flame spewing out of the ground-floor front bay window. That is a very bad omen at this hour, with residents asleep on higher floors unaware of the rising peril of super-heated smoke and toxic gases.

The speed at which the rolling, rapidly rising brown trail of smoke is billowing from the uppermost front windows is another very bad sign. It indicates a very hot blaze, possibly deliberately ignited at this mild time of year. In the cold winter months, we regularly see fires of this severity caused by paraffin heaters. But it is not cold enough for those, so a real evil is sensed as we throw out hose like lightning, and slip and pitch the escape in seconds, ready to carry out any rescues needed from the front windows.

But no persons are visible there, and Jack Hobbes has already accessed the rear by knocking up the neighbours and going through that house to access the rear of the fire building. There is no one at the rear windows either, but heavy smoke is issuing and we can now see that the pressurised thick smoke is also percolating from the roof slates, indicating much heat and pressure contained. This is a real nasty fire all right and, as Ricky Tewin and I – the BA wearers tonight – rig as quickly as we can and prepare to make an entry via the top-floor front window, the dreadful smoke is pouring out as thick as a bale of wool.

Others have got a line of large diameter hose to work. Once the front door is forced, the fire can be seen racing up the staircase as if it were a chimney flue; but the two-and-three-quarter-inch hose will deliver rapidly the water necessary to quell this evil blaze. A rapid fire suppression is vital to giving whoever might be inside a better chance of survival, and it improves our prospects of coming out in one piece. We are going into the highest and therefore hottest part of the fire. If the crew with the jet can quickly quell it on the stairway, we can hopefully search each floor, gradually getting to the relatively cooler part of the building lower down. But a fire of this ferocity can soon burn away a timber stair, or if the stairs are made of stone and hit by a cooling jet, crack and collapse, plunging crews into the inferno beneath. That awful thought that 'this could be the night' lies in the subconscious, but still our heartbeats increase and the adrenalin surges.

Biff has called for three more Pumps, as the three crews now on scene (our two and the one from down the road) are fully stretched. From the loudspeaker on the Pump's radio, the metallic-sounding voice of the operator at Fire Control can be heard repeating the assistance and informative message just sent by Biff: '*Make Pumps Six – Persons Reported. It is a house of four floors, 15 by 20 feet; 50 per cent of ground floor and staircase from ground to top floor alight, remainder of building smoke-logged, unknown number of persons involved. Escape, BA and one jet in use.*'

The blue beacon atop a white car signals the arrival of some 'brass'.

It is Ernie Tappisson, the ADO from Divisional HQ. Tappisson is a sparely built man of medium height with an expressionless face. He is a disciplinarian in the mould of many of the top brass within the fire service, but he is not vindictive. Long years spent in the brigade's busiest locations and his military service have simply convinced him that urgent life-endangering situations require no-nonsense positive leadership if lives are to be saved and the safety of crews looked after as much as possible.

He walks briskly over to the Breathing Apparatus Control Vehicle, where Biff was just formulating an informative message, and waits while the Staff Sub Officer finishes writing the text of the message down.

'Good Morning, Mr Sands. What have we got?' Tappisson asks.

The Station Officer gives him the situation report, outlining his view that the way that the fire had spread so fast probably indicated deliberate ignition and the possible use of accelerants.

The fire service has a command structure linked to the nature and size of the incident. As the number of Pumps and other appliances increases, the rank of the officer in charge increases also. But the ADO knows and respects the competence of Biff Sands – although he stresses that he is remaining at the incident, he does not take over the command.

I am now at the head of the escape ladder, some 40 feet above the street. The smoke from the windows is still coming out at a rate of knots, indicating the great heat still within the fire. It is not every day that we get house fires that burn as fiercely as this one, but I have been to others of a similar size, so I know the drill. I carefully push the back of my hand into the rolling brown smoke. It's very hot, but I don't feel the skin tighten, as it would if it were so hot that the whole room was about to explode.

I look down the escape. Ricky is standing just a few feet below and when I give him the thumbs-up, the BA mouthpiece gagging my speech, he knows that this means we are going in. I am reassured by the sight of the charged line going into the front hallway. The quicker that the guys on the jet knock down the heavy body of fire, the better and safer will be our environment. In addition, the survival prospects of anyone inside will also increase.

The flames and heat had caused the window glazing to blow out ten minutes earlier, so I straddle the sill, taking care not to get cut on the few spikes of glass still left in the sash frame. I grasp the small, thin rope line that is attached to the head of the ladder; this will assist us to relocate it if we have to make an emergency exit. I find the floor of the room with my boot and drop down as low as I can, and even there, only a few feet lower, I feel the temperature decrease a tiny bit, which spurs me on.

Visibility is zero in this super-heated fast-rushing fog and it is only by touch that I know Ricky is in there with me. My goggles have already misted over from the sweat from my head, but what's the difference? In this sort of scenario, from now on everything will be done by touch, and hand claps and oral grunts.

I keep a tight hold on the line and Ricky grasps the rear of my harness, and together we systematically feel our way all around and across the room. There are what feel like two single beds along a wall. We search on top – no one there. Then underneath as best we can, hampered in our movements by the bulk of the chest bag of the BA. No one there, either. I can hear the crashing of plaster and broken wall and ceiling laths, as the powerful jet working its way up the hot chimney of the staircase does its vital job of fire suppression.

Reinforcing crews have taken a second large jet into the house via a ladder pitched to the top floor at the rear and these will be working towards each other, even as we search.

As far as we can make out in this zero visibility, there are no persons in the front top-floor room, so I release my tight grip on the 20-foot line, surrendering the security it provides, and turn right out of the bedroom, crawling along what must be the landing. Immediately, my hand sinks into the palpable mound of a person's stomach. I carefully move my hand upwards over the belly and onto what feels like the heavy bosom of a large woman. My hand touches the hot perspiring skin of a face, recognisable by the nose and mouth. Most of the horrible flames on the staircase have been quelled and the jet men sound very close.

I grunt out to Ricky, grabbing his hand and pulling it onto the body, so that he is aware that we have located a victim. In this sort of fire and heat, and at this time of night, it is possible the victim has been unconscious from the smoke for a long time and is probably already dead. But we cannot know this.

Instinctively, I foresee that, given the victim's size and dead-weight, getting her onto the escape, hampered by the BA set's bulk, will be very difficult. If we can get her up to the window and get her head into the air, we will have improved her survival chances, but what resuscitation can be applied in that location in this deadly smoke?

Because the large jet has knocked down a good deal of flame, the stair seems now much friendlier – no longer the raging chimney it was on our arrival – so I decide to grab the victim under the armpits and, with Ricky's help, drag her backwards down the stairs to the street, where resuscitation at least can be given.

The smoke is much cooler and so I chance my eyeballs, lifting up my goggles. Wonder of wonders, I can see Ricky. It turns out that as thick as the smoke has been, my misted-up goggle lenses have not helped my vision. He lifts his own goggles and I indicate we are heading down and out.

I can now see that the woman is about 5 ft 8 in. and maybe 14 stone. I swivel her bulk across the top landing as a BA man with a jet comes up the top flight. He sees us and shuts off the hand-controlled nozzle, lays it down and assists us in the long bumping and heaving rescue. If not already dead, the cuts and bruises of being dragged 50 feet around a dog-legged stair full of fire debris will be a small price to pay for survival.

Unfortunately, this woman does not survive the fire, nor do two other occupants – all three victims of a never identified thwarted lover who used a can of petrol poured through the letterbox to extract his or her revenge.

It is a tragic and horrific end to life.

Chapter 19

London's Arteries

It is our second day duty and we are just munching the traditional London Fire Brigade mid-morning snack of shredded cheddar cheese inside a couple of buttered bread rolls.

'Get thee behind me, Satan!' exclaims John Lydus, a keen cross-country runner who has won a cupboard-full of trophies, as I jokingly tempt him with one of the huge rolls.

'If you knew what strong cheese does to your arteries, you would never touch it again,' he says, shaking his blond head.

'But surely, John, a bloke who runs over 50 miles a week must burn it off, yeah?' asks Jerry Pashleigh, as he starts on his third roll.

'Well, I suppose the miles I run must offset some of the effects of the fat in dairy products, but give me a bowl of tuna every time. I'd like to see 40 by not clogging my arteries with gunge,' he replies.

'Each to their own,' interjects Paddy Mulligan with a grin, which suggests he's in a mood to stir up a debate. 'One man's meat is another man's poison. It's the same with exercise, John. You love your running, whereas I believe that the big fella above gives us all so many heartbeats when we're born and who's to say that you might have used up most of yours already, pal?'

'You might be right, Paddy—'

But the clarion call of the turnout bell cuts off John's response and we all jump up, some with mouths still full, and pour down the pole.

The green light indicates the call is for the Pump only and the duty man confirms it with a bellow, '*Pump only, 22s ground*', handing the guvnor a slip on which is printed, '*Sewer man failed to surface, High Road, opposite the infirmary*'.

As I clamber aboard with John Lydus, I cannot help but smile as I think how apt is the 'shout' we are responding to. In an article I had recently read about Joseph Bazalgette, the inventor of the London sewer system, the author had likened it to the 'capital's arteries'. I hoped that if this call saw us descending into them, they were not as blocked as John thought his own arteries might be if he indulged in the fire station cheese rolls.

As well as attendance from the local station, the ordering slip shows that the ET (Emergency Tender) from Euston is ordered, which is routine for any report of this kind. The men who work in the London sewers are exposed to a number of hazards that can be life threatening. For example, a sudden heavy rainstorm can cause masses of water to enter street drains and rapidly fill sections where crews are involved in maintenance and inspection tasks. For this reason, a 'top man' is positioned at street level to keep an eye on such weather changes.

Another potentially deadly situation is created when foul or toxic gases are present, either from chemical changes within the effluent or as a result of someone illegally discharging harmful materials into the sewerage system.

As we weave in and out of heavy mid-morning traffic, I know that our Station Officer will be running over in his head the standard emergency procedures for this type of incident. He will be recalling the need to locate the top man, and to ensure that manhole covers are lifted in the street to allow some light into the chamber and let foul air out.

'Get your BA on, ready to start up, just in case,' Biff turns and shouts to us over the din of the horns and engine.

John and I take off our helmets and pull the mouthpiece-securing harness over our heads. We get into the set's bulky harness, not easy when swaying about at 40 miles an hour, and tighten the body belts. Although the incident we are racing towards is on another station's area, our boss won't know until we arrive if there is a Station Officer riding in charge of their appliances. If there is, then he will be the officer in charge. If it is a Sub Officer or Leading Fireman, then our guvnor will hold command responsibility.

As it is, we can see the beacons of appliances 50 yards ahead and the white helmet of a Station Officer, so Biff can concentrate on our supporting role.

Barry Priestley pulls our Pump up behind the sewer men's van and we all dismount, awaiting instructions from Biff, who is soon liaising with the other guvnor and two sewer men identifiable by their helmets and safety harnesses.

'Harry and Pete went down to do a routine inspection,' one of the men is breathlessly blurting out, clearly in a state of distress. The agitated young man goes on: 'They went down the chamber just opposite the hospital entrance and were due to surface up the chamber adjacent to the newsagent's 100 yards north. Pete came up coughing and gasping. He said that they had been hit by foul air and there was no warning on their detectors. Pete said Harry was behind him, but he ain't surfaced.'

A red-faced man with a harness, presumably Pete, is on the ground, his back leaning on a telegraph pole. Another man is giving him oxygen from a cylinder.

Biff runs over to John and me. 'Look, I know that the ET should be with us soon with a specialist BA crew, but we can't afford to wait. So start up now. Take a long line and an oxygen cylinder, in case you locate the missing bloke. Lay out a guideline and head straight up from here,' he instructs calmly and positively, and with the coolness for which he is renowned when the chips are down.

Within seconds, mouthpieces are in, nose clips are secured, main valves are open and the BA tallies are given to Barry Priestley, who has already secured one end of the guideline to a lamp standard nearby.

Reassured by the distant horns of the ET, we kneel down and, looking down the chamber, locate the iron steps that are fixed into the shaft's wet side walls, our boot soles echoing. The daylight coming down the shaft helps us locate the steps as we descend the 30 or so feet into the sewer itself. The guideline is secured to me inside a red canvas bag and it unwinds as I descend. Once I am at the bottom, I look up and can see John's chest-mounted lamp giving off its barely adequate light. The water and effluent is about a foot deep and it is slippery underfoot, which is why the ET crew wear specially soled boots with studs to give a better grip.

I look along the arched sewer, which is about ten to fifteen feet high, and can just make out channels every so often in the side walls, which I know lead to the drains and gullies above. I can also see the shafts of daylight from the opened manholes far ahead. I see an iron stanchion on the wall to my right and tie off the guideline before proceeding slowly. John is about ten feet behind me and we move from side to side, feeling for the missing man with our boots.

After about five minutes, we are close to the foot of the shaft from which the other bloke surfaced. The effluent feels thicker here and my lamplight illuminates bubbles on the surface of the thick sludge, now made more visible by the daylight from the shaft just ahead. I move gingerly through the black gunge.

My heart races as I notice a lot of bubbles in the turgid liquid. This is probably poisonous gas and I instinctively place my left hand on the outside of the mouthpiece, gently ensuring my nose clips are properly in place and checking the tightness of the oxygen seal, just in case.

Then, illuminated by the rays of daylight from an opened manhole far above, I can see a bulky form half-submerged against the bottom of the shaft. I clap my hands to alert John and wave my lamp, and in a second he is at my side. We can now see a red helmet and the slumped form of the missing man. He appears unconscious, probably overcome by the fumes. For him, it was too late to get into the shaft and escape, as his mate had. At least his head is out of the water and effluent.

He is in a half-sitting position, leaning back against the side wall three feet from the first iron rungs up to safety. John has the black-and-white spare oxygen cylinder. He cracks its wheeled valve and lets the pure oxygen seep out into the victim's nostrils. I signal to John to climb up the shaft to the street. We know the ET crew will probably be there by now and will be able to send their crew down with a rescue sling or Neil Robertson stretcher and a portable resuscitation set.

It is now that John's high physical fitness, gained from his cross-country running – and perhaps his cheddar cheese aversion – will benefit him, as he begins the arduous vertical ascent, encumbered by the 35 lb BA set.

I continue to hold the oxygen valve to the man's face and feel for the carotid artery pulse. If there is one, I cannot detect it and I fear that we are too late.

Within little more than a minute, I can hear and see the ET crew descending. An oxygen mask from the portable resuscitator set is strapped to the sewer man's head, then he is tightened into a rescue harness, attached to a hoist erected far above, where traffic trundles along unaware of the life-or-death drama being played out below them. Within ten minutes, he is out and on his way to the infirmary, which fate has decreed is only yards away.

The super-fitness of John Lydus, the fortunate closeness of the infirmary and the almost immediate attention administered helps ensure his survival, which means we have notched up another satisfying lifesaving response, but from a location in which, for once, fire and smoke are absent; instead it is the capital's effluent and the deadly gases that are sometimes produced by it as it flows through the subterranean arteries of Mr Bazalgette's famous sewer system that have caused us concern.

Chapter 20

Perilous Paraffin

Within London during the later years of the twentieth century, especially in areas with a lot of large Victorian houses separated into flats, or older tenement-type dwellings, the brigade witnessed many serious fires caused by paraffin heaters and these regularly resulted in death and injury.

During this period, such stations as North Kensington, Brixton, Harlesden, Paddington, Holloway and Islington responded regularly to fires in these kinds of premises, often in the socially deprived parts of London's boroughs and home to the capital's immigrant populations.

In the often icy winter months, especially in properties without central heating, it was customary to make use of the portable paraffin heater to try to counter the cold – a cold that would be more acutely felt by the elderly population in general and specifically those from foreign shores unaccustomed to freezing temperatures. Paraffin was a cheap form of heating and was readily available from dispensing pumps at the 'open all hours' corner shop or local petrol station. Any portable heater has an inherent fire-risk potential, but this risk is magnified when flammable liquids are used as fuel, particularly when no automatic cut-off to the supply is provided, which was the case until regulations were introduced to remedy this potentially lethal omission.

The shrill teleprinter indication bell has barely begun its ring when it is followed by the much louder urgent clamour of the call bells and the automatic switching on of the lights, bathing the semi-darkened station in their yellow glow. It is just after 1.30 a.m. on a freezing cold Saturday in January. Up to this point, the night duty has been relatively uneventful.

At twenty minutes after seven, the call bells had tolled, sending us to '*Persons shut in lift*' in the tower block just a few streets away – something that always seemed to happen more often over the weekend. Difficulty in gaining entry to the lift motor room had turned a 20-minute incident into one taking almost an hour.

When we returned, Geoff Joynt, the 'chef' on this watch, was serving up a spicy Indian curry to stave off the cold. Geoff is a man of medium build, in his early 30s, with a deadpan expression that belies his creative mind and his skills as a photographer – to supplement his fire department salary, he makes use of his talents with the camera at weddings, a ceremony he will soon be taking part in himself after his long-term fiancée finally agreed to become his wife.

'It sure smells great, Geoff,' said Andy Carlisle, licking his lips in anticipation and forking up a chunk of highly spiced chicken breast, then blowing out air as the high-octane curry powder met his taste buds. Andy was just about to comment on the spice when the harsh ringing of the bell sounded and another call came in. We all leapt up and headed towards the pole drop.

'*Fire in rubbish skip – Pump only*,' shouted out the duty man for the shift, Niall Pointer. At least the Pump Escape's crew could enjoy the curry, I thought, as my feet thumped onto the rubber mat at the foot of the sliding pole.

With a roar and hee-haw of siren, we charged out into the already icy night air to travel the half-mile to the burning skip. It was only a small fire on the surface of waste deposited by builders renovating a nearby house and in ten minutes the hose-reel tubing had done its job and we'd returned, via the 'back doubles' of shabby down-at-heel streets, to the station to polish off the curry, followed by some cooling vanilla ice cream.

So as the call-out bell sounds at 1.30 a.m., two of the watch are lying on their bunks, the atmosphere fuggy from the heat of the radiators; three are watching the end of the midnight movie; one is studying for a forthcoming promotion exam, while the rest are sitting in the darkened watch room, positioned at the front of the station, adjacent to a main traffic route, watching drunken revellers slewing their way homewards from the many pubs, clubs and dance halls of the area.

The watch room is suddenly illuminated as the bells clamour and the founts of the teleprinter begin their metallic chatter, darting quickly on their carriage across the teleprinter roll and then springing back. '*Fire in house – multiple calls being received*' is the cryptic message that sees our two Pumps, plus the next nearest Pump from an adjacent station, dispatched.

Multiple calls means that several 999 calls have been received, which results in Fire Control automatically dispatching a fourth appliance. Multiple calls also indicate that this could be a working fire; the early hour, along with the sub-zero temperatures, does not bode well, especially as the street to which we have been called has become infamous over the years for its serious house fires.

Right out of the station, left at the traffic lights, right again, third right . . . the drivers could almost get there blindfolded, so familiar is this location: a gradually sloping avenue of Victorian redbrick terraced houses from three to five storeys in height, all with stone steps leading up to a front door, some with dormer attics 50 feet above the ground, others with a basement.

I can feel the bitter cold of the morning quickly numbing my exposed left arm, next to the half-open rear crew cab window, before I slide it into the cloth sleeve, still slightly damp from the water used on that earlier skip fire.

The lead Pump Escape, by tradition the principal lifesaving ladder, and always first away from the station, makes an arcing turn into the street, causing a gallon or so of water from the onboard tank to cascade onto the freezing road; it will soon turn to ice. The Pump follows suit, and we see that a browny-grey fog of smoke has already blotted out many of the street lamps; in the air is that pungent and unmistakeable odour of burning wood, plaster, rubber, fabric and paint. Any doubts we might have had that this was a serious fire are rapidly removed when we notice, to our left, angry orange, crimson and yellow flames leaping outwards from windows on the ground, first and second floors of a four-storey terrace. The ruddiness of their colour is reflected on the thick pall of brown smoke.

'Stone me! It's going like a torch. Have we got a goer here or what?' shouts out Andy Carlisle above the blasts of the two-tone and the roar of the diesel engine. Even at such an early hour, and in spite of the cold, there is a small crowd of onlookers, mostly neighbours, some with overcoats pulled over their nightclothes, other folk returning from a night on the town. Some are pointing upwards. Several of the women have their hands flattened over their mouths in shock, having noticed, through the smoky haze, three people trapped by this fierce blaze, waving frantically and screaming for help at a second-floor window.

The Pump Escape squeals to a halt about 40 feet past the house to allow room for the huge ladder to be slipped and positioned. Like on so many similar streets throughout this huge city, cars, vans and small trucks are parked nose to tail, such dense parking invariably causing problems when positioning ladders during firefighting and rescue operations. The Pump, meanwhile, shudders to a stop some 40 feet to its rear.

'Slip and pitch to the second floor, and get a covering jet to work pronto,' shouts Station Officer Ben Tuke, the bulk of his tall frame and 17-stone bodyweight accentuated by the comb atop his white helmet and the drawing-in effect of his belt, which exaggerates his barn-door shoulders even further.

The rescue procedure ordered is a bread-and-butter action for inner London stations and is practised almost every day, meaning it can be done automatically. Similar to the way in which a soldier will provide covering fire whilst others advance across an exposed position, the 'covering jet' is a powerful water jet and spray from a large hose. It provides some protection from the flames, smoke and heat below rescuers and the rescued who are on a ladder above the fire; it also assists in providing a more breathable atmosphere.

As the escape is pitched, others grab delivery hose from lockers and sling it along the road, where it rapidly unrolls with a sharp slap against the icy tarmac.

To the coupling at its end, another man connects a nozzle to provide the high-velocity covering jet. Geoff Joynt, the Pump Escape's driver, has connected the other end of the hose coupling into the delivery of the side-mounted pump. He spins open the wheeled valve to send the 300 gallons in the onboard tank surging along, fattening the flat hose line, and emerging with a throaty hiss and crack, looking like a silver lance, that rams into the wall before cascading from the hot redbrick heated by the searing flames.

It is at such scenarios as this, with the congestion of parked vehicles obstructing our closer approach, that the almost one-ton escape comes into its own. Its ability, because of this weight and stability, to be positioned over the tops of parked cars and over front entrances is such that it is one of the most effective rescue ladders around, its solidity making it almost like a staircase.

Jim Pile is extending the escape like a man possessed. Within half a minute its head is resting on the hot wall in the lee of the flame and smoke, just to the side of the second-floor window, where two women and a small boy are trapped; the main stairway has been rendered impassable by the heavy body of fire roaring up it as if it were a chimney. Almost before the top section of the ladder halts, Andy Carlisle is ascending the rungs like a circus monkey running up a pole.

Andy, whose rugged good looks have earned him the nickname 'Steve McQueen', is a single man of 33 who makes no secret of his liking of women. He is no Flash Harry, though, and his quiet demeanour belies his ability to attract female attention. He keeps fit by swimming and weight training, and is one of the most respected firemen in the district, let alone the station. If there is anybody equipped to excel in the tough world of sharp-end firefighting and rescue, it is him.

I have laid out a second line of hose from the Pump, ready to start knocking the fire down at ground level, whilst the BA wearers rig. I inwardly praise Geoff for already having connected up to a street hydrant to provide a limitless supply of water, knowing that the onboard supply lasts only a few minutes when using large diameter hose lines. As the water speeds along towards me, crouched low outside the front entrance, I can hear the loudspeaker of the Pump's radio repeating the guvnor's message – '*Received from Station Officer Tuke: Make Pumps Four – Persons Reported*' – and in the distance the sirens of reinforcing appliances.

The request for the extra pump is academic; every call to this high-risk locale guarantees three pumps, but the repeated 999 calls received means that Control automatically dispatch an additional pump, which will be well en route by now. The receipt of 'persons reported' will see an ambulance on its way, too.

During my visit to the New York Fire Department, I had noted some of the differences between radio messages sent from fires by the New York officer in charge compared with our own in London. One that I especially remember was the transmission '*All hands working*'. This informed chiefs and men in surrounding firehouses that those in attendance were all fully engaged with fire suppression, ventilation and/or rescues. In the City of New York, such a transmission was often the prelude to a serious request for extra appliances and crews; it was a signal to firefighters nearby that soon they might be rolling, too.

'All hands' are certainly working here, and the two other crews, now speeding up the slope towards us, will be vital to our endeavours.

'Sub, get yourself around the back, and check the situation and report back to me,' shouts Station Officer Tuke to the Sub Officer just arriving on the first reinforcing Pump. Although events at the front of the building on fire are usually obvious, those at the rear can be easily overlooked in the initial frenetic activity at a rescue job like this one. Terraced properties create their own problems for accessing the rear and if the body of fire is such that no access to the back is possible, then the quickest way there is through the property next door.

This morning it is a simple matter to get to the rear of the affected house via the immediate neighbours, who are themselves out in the street, fearing their own house will be lost. It is an understandable fear, but most houses in London of this construction were required to be built with a party wall that extends above the ridge of the roof, and it is rare for even the most severe house fire to spread to homes on either side. But, for now, the situation at the rear isn't my concern. Our efforts are concentrated on knocking down the severe fire at ground floor.

With Jim Peating's short but heavy frame humping the hose behind me, I direct the jet firstly to the ceiling of the hallway, then move it in a continuous sweep. We crouch low as the powerful jet brings down light fittings, and wall and ceiling plaster whose bond has been weakened by the heat of the fire. The jet creates a cloud of skin-reddening steam as the water absorbs the heat of this inferno. It is a raging mass of orange, crimson and yellow veiled by a semi-transparent fog of acrid brown smoke that seems to sear into the eyeballs, the tear ducts running instantly in a bid to protect them.

Fortunately for us, who are not protected by breathing apparatus, most of the smoke is rapidly venting out of the windows, whose glazing has been shattered by the heat and rolling flames. Even crouched so low that we are almost prone, the heat is intense. I can feel rivulets of perspiration running down my back and legs. I wince as a piece of hot ember finds its way past my neckerchief onto the back of my shoulders, rapidly blistering the skin.

I swirl the nozzle around and aim the water stream at the ceiling of the stair landing, then at the walls of the stairway. A door to a room off the stairs is open, so I direct the jet onto the ceiling of that room and, inching slowly forwards, with Jim humping the heavy hose line, rotate the jet round and round, soaking the room's walls and lower levels. The jet has always to be kept in motion. This enables as much of the burning materials as possible to be extinguished via the actual knockdown force of the stream at the same time as cooling the burning mass. Such is the heat at fires like this that a flame will be knocked down only to leap back as soon as the jet is taken away. It is therefore a running battle of water versus the fire triangle of fuel, oxygen and heat. This is why it is called firefighting.

After a few minutes, we have cautiously ascended to the first-floor landing; with each extra foot gained in height, the hotter become our surroundings. We move tentatively all the while, in case the severe fire has weakened the staircase to the point of failure and collapse.

As we gingerly progress ever upwards, hauling up the heavy hose line, I cannot help but recall a photograph I saw years earlier. It was in a book about the history of the London Fire Brigade and, in its own small way, it had helped inspire me to become a part of the capital's firefighting force. For some reason, it had lodged itself in my mind. It was a black-and-white image of two firemen slumped against a wall with their helmets removed. Their faces were white, shocked, soot-stained and grimacing with pain. It was captioned 'Showing the firemen injured when a staircase collapsed during firefighting operations at Upper Thames Street'.

Amidst the noise of the fire and the hissing and crashing created by the jet, we can feel the staircase slightly vibrating. I begin to worry that we too will end up in that same predicament, though I need not have done: I soon realise that the vibration is caused by the bulk of Station Officer Ben Tuke ascending. His huge form is suddenly crouching at my shoulder.

'How are you doing, men?' he asks.

'OK, but it's bloody hot, isn't it?' I reply.

'Certainly is, son, which is why a couple of BA men are about to relieve the pair of you. You've done well to get this far in this heat. The ADO's come on and he's made Pumps six and ordered a set of ladders [the 100-foot Turntable Ladder].'

'OK, guv, we'll keep pushing upwards until relieved. How many persons are involved?'

'Not sure, but neighbours say the house is divided into flats and bedsits. Could be ten or twenty or more,' he replies.

Three miles away, ADO Tom Monsal had been using the early and quiet hours to complete a report when Control advised him that repeat calls were being received for this fire. He switched on the receiver monitor, which allowed him to listen to all radio traffic across the whole brigade area, and within a few minutes had intercepted the '*Make Pumps Four – Persons Reported*' message.

He had been a Station Officer himself in the area of this fire and knew from first-hand experience the number of serious, sometimes fatal jobs that had occurred here all too often over the years.

It is brigade policy for the duty ADO to go to all 'persons reported' jobs, but Tom Monsal has been around a long time and knows that too prompt an attendance can cause those officers below him to think that they are not trusted. Accordingly, he decides to wait to hear the first informative message from the incident.

He doesn't have to wait more than a couple of minutes: '*Terraced house of four floors, approximately 30 feet (width) by 50 feet (depth), 75 per cent of ground floor, 50 per cent of first and second floors, and staircase from ground to top floor alight, unknown number of persons involved, building being searched by BA crews, Escape Ladder and covering jet in use.*'

The message paints a clear picture of a serious fire and one that could develop further; he has no doubt that it is time to go on and take overall responsibility for operations. Within minutes, his staff car is mobile. He is on scene in ten minutes. After being briefed by Ben Tuke, he formally takes over command. His lengthy experience enables his decisions to be informed and soundly executed, and soon he has sent his message requesting extra pumps and the Turntable Ladder.

Although a party wall at ridge level would help prevent spread to adjacent houses were the fire to burn through the roof, it is not unknown for persons trapped above such a severe fire to make their way to the attic and try to break through the slates, or a skylight, in a frantic effort to escape from the heat and smoke, then to get onto the roof and hang perilously over the slope, hoping to be rescued. Were this to happen here, then at least there will be a tall enough ladder on scene to better facilitate a rescue.

Although it seems like ages, Jim and I have only been inside the house for 15 minutes before being relieved by the BA crew. We could have pressed on, but why beat up our bodies unnecessarily, having already inhaled enough heat and smoke to weaken us, when the BA-protected team can ascend to the highest and hottest levels of the house with relative impunity?

After the sauna-like heat of the staircase, going back out into the sub-zero air is like taking a plunge into an icy pool, but the sharpness of the night is recuperative and we stop to look back at the scene, our perspiring bodies steaming like two horses who have just finished the Grand National.

The ambulance, automatically mobilised on receipt of the 'persons reported' message, is parked in the centre of the street, its rear doors flung back. Its stark interior light reveals three individuals in the back, draped in the official red blankets of the London Ambulance Service, their faces covered in an oily grime, mixed with that clammy perspiration that is the hallmark of shock and relief.

Later, at the post-incident debrief, Andy Carlisle told us about his involvement in the rescue of the three of them, as did Jim Pile.

Andy didn't have time to worry about the fire and smoke belching out of the windows as he put into life-or-death reality a drill evolution that is familiar to the inner-city crews. Slipping and pitching the escape ladder and getting a covering jet to work is a basic manoeuvre taught from the earliest weeks of training and polished to perfection thereafter.

It wasn't the first time that he had been in this position of rescue, and he remembered the first time well. It had been a severe blaze, in which three had died in a similar house to this. Before that particular incident, he had often tried to imagine what it must be like to be trapped in a burning building with your escape route shut off by searing flame and poisonous smoke, especially in the small hours, when few people are around to hear your cries for help, even in the centre of London, one of the biggest cities in the world.

How welcome, he thought, must the distant note of horns be, almost calling out to those trapped to 'Hold on, hold on.' How welcome the throaty roar of engines and the sight of those spinning blue beacons, heralding the salvation surely prayed for by those imprisoned in the skin-blistering flames and heat, held back only by the room's timber floor and flimsy, ill-fitting door.

He was at the head of the ladder in seconds and locked himself onto it by pushing a foot and leg through one rung and then wedging it under the rung below. Through the smoke and steam coming up from the belching fire below his position, he had seen that two of the occupants of the room were women, probably in their 50s or 60s, and the other a boy of about 12 or 13.

One woman was plump, clad only in a green cotton nightdress.

The other was noticeably skinny, even with the bulkiness of a towelling housecoat. The boy had on a pair of red pyjamas with a Dennis the Menace print on the front. All of their faces, but especially the small boy's, had a transfixed look of petrified fear.

They were straining their heads and upper bodies out of the open sash window in an effort to escape the pungent brown smoke that had seeped under the door and through gaps in the timber floorboards. Sinister smoke was now harassing and intimidating them with its cloying, clinging presence.

Andy knew that the floor might be weakened by the intense fire below, so much so that it might collapse at any time. And that the bedroom door, their last line of defence from the fire and heat, might be about to give, consuming the room and its three victims in a flash of exploding flame. There was no time to lose.

'Right, I'm here and will get you all down, but do everything I say,' he shouted above the crackle and crash of the blaze, and the drone of the fire pump some 30 feet below.

The ladder vibrated with the thud of boots.

'Andy, I'm right below you, mate. Pass 'em to me when you're ready,' called Jim Pile, who had a few minutes earlier hauled the ladder to its present position. He had now mounted it and ascended to a few rungs below Andy to provide those vital extra hands needed to safely bring the three down.

'What's your name, old son?' Andy asked the shivering and petrified boy.

'Peter,' he half-sobbed.

'Right, Peter, you'll be down in a jiffy. Just stay there, I'm coming in.' With that, he released his leg lock and, in a bound, was in the furnace of the room. He could see the thick smoke pushing up around the skirting boards. It looked just like a slowly moving roll of hot, brown cotton wool. With each second, it grew more dense, and the timber floor felt springy, the tongues of searing flame below curling around and into the ceiling laths, floor joists and old floorboards of this Victorian-era house.

Jim was now at the ladder's head and in a flash Andy had picked up the small boy. 'Don't worry, Peter. I'm passing you to Jim on the ladder. Hang on to him tight and do as he says, right?'

Jim's outstretched hands and arms gripped the boy with all the strength he could muster. For a fleeting few seconds, he had pictured his own 14-year-old son, now safely tucked up in his bed in the family home in east London, and held on to the scared little boy even tighter.

'Put your arms around me, son, grip tight and don't let go until we are down, yeah?' he said.

The boy grunted OK; his own instinct to survive ensured he did what he was told. In a minute, Jim had descended sure-footedly, in spite of the heat and smoke issuing from below, and, grateful for the spray from the covering jet, handed the young lad down to outstretched hands at the heel of the ladder.

Thirty feet up on the second floor, Andy had been inwardly concerned at the fire-weakened floor on which much of their fate depended, but he couldn't show it. Suddenly, the woman in the housecoat let out a shriek. 'Oh God, the fire's coming through the floor.' Sure enough, red flames, tinged with yellow, were licking six inches or so up on a far corner.

'It's going to be all right, we'll be out in a tick,' Andy said. 'Trust me.'

He had moved a yard to his right and was heartened to see Jim ascending at a run. 'Jim, be ready. I'm going to put them out one after the other, the floor's iffy and we've little time,' he shouted.

'Ready when you are,' Jim coughed back, a pungent wave of smoke rolling out of the window and catching his breath.

Andy had got both women to the right-hand side of the window near the white-painted head of the ladder. The plumpest one would be the first out – he reckoned he could grab the skinny one by her waist while holding onto the head of the ladder if it had come to it, in an effort to bale them out.

'Oh, Holy Mother of Mary, help me,' the plump woman cried out in an Irish brogue as broad as the River Shannon, as she looked down at the hard pavement below and saw the fire, smoke and steam.

'Get hold of the ladder here,' Jim said, gripping her plump bare left upper arm to steady her as she did, his strong fingers sinking into the doughy white goose-pimpled flesh.

'Now, get your leg over the sill and put your foot on the ladder – I'll guide you.' Andy secured her from the room side by her right arm and, with a pull, she was on the ladder.

'I'm right behind you, love. Just keep your hands holding on tight and I'll guide your feet. Take it easy, nice and easy,' reassured Jim, noticing the nervous tremor in her calf as she stepped down.

'Oh my God, oh my God, save me, save me,' she sobbed. Jim was directly below her small slipper-clad feet and, as the combination of fear and relief hit her nervous system, a stream of hot urine hit him across the face. Then she was down, and covered with a red blanket by the ambulance attendant, and assisted to the waiting vehicle.

Andy, in spite of the critical nature of the situation, had managed to establish that the skinny woman was called Philomena. In those tense moments, he had noticed how developed the fire had become under the floor and had got her to sit astride the sill where he had positioned himself, his left arm around her and his right hand gripping the ladder's head.

Should the burning floor fail before Jim came back up, he could grab her and knew he had the strength and power to hold her and pull them both onto the ladder. He later said it felt as if he had been in the room for ages, but no more than five or so minutes had elapsed since he had first run up the ladder to the three trapped inside.

Breathing hard now with the exertion of carrying down the boy and assisting the first woman, Jim's sooty, black helmet emerged again from the smoky atmosphere. In a few seconds, Philomena was slowly being guided down to the ground.

As Andy mounted the ladder himself, the flames had burned through about six square feet of floor. Even though the jet hose now at work on the ground floor had begun to knock down the fire, it was too late to prevent the floor from collapsing. It did so with a thump and a crash, the hot draught sending a torrent of smoke and sparks up the front of the house.

'That was a bleedin' close shave,' Andy thought, preparing to descend, reflecting on how very differently things might have turned out had we arrived a few minutes later.

'Ace piece of work there, Andy and Jim,' Ben Tuke shouts, as I fill my lungs with the icy air. 'A first-class rescue above a real bastard of a fire.'

'Cheers, guv, but it's what we're paid for, ain't it? Although it don't half prove the value of regular drilling on the basics,' Andy replies.

After Jim and I are relieved, we go round to the rear of the house via next door to see if any help is needed.

Large flames had been curling out of the top-floor windows, but these are quickly dying down and the grey smoke is turning to steam, as the BA men who had relieved us start to knock down the remaining fire with the large diameter hose, its huge gallonage rapidly absorbing heat and cooling red-hot walls, floors and ceilings.

We saw no occupants shouting for help from the rear, but that didn't prove no one was there. There could be an individual now lying unconscious below the window, behind a door, on or under a bed or settee, prostrate on a staircase, landing or hallway, or overcome in an attic space, having tried in desperation to break through the roof to escape the fiery, malevolent monster beneath and around them.

We return to the street to report back to Station Officer Tuke. As we come opposite the front entrance, the heavy thump and scrape of fireboots and a dragging sound comes from the staircase. Two BA men, part of the reinforcing crews who had been detailed to make a floor-by-floor search for the persons reported, emerge from the smoking, steaming and fire-charred hallway. They are pulling by his armpits and on his back down the remaining few stairs a heavily built male. He looks about 15 stone and would stand at six foot surely, clad only in a string vest and Y-fronts; but he carries the dead weight of unconsciousness.

A BA crew located him behind the door of a bedsit on the top floor. There was no time for the niceties of rescue techniques, as they allowed his weight to assist the humping slide down the now passable stairway, dragging him over the still smouldering spalled plaster and charred wood and floor coverings. Any bruising or lacerations from this is the price to pay for a quick retrieval to the outside air – it could help save him from death.

The first can be repaired, whilst the second is irreversible.

We rush up to grab an arm and a leg to assist the BA men, spent from bringing this heavy victim down 50 feet of stairs. We place him on the freezing pavement, which contrasts with his skin, hot and perspiring from the heat of the fire.

Niall Pointer's excellent work with the covering jet is now over and in a flash he is at the man's side, his thin fingers trying to locate a pulse in the folds of flesh in the man's bull neck.

'Can't feel anything. I think his lips are blue,' he says, shining his lamp onto the man's grey complexion.

A green canvas salvage sheet appears from thin air and, in a second, the strong arms of four firemen have lifted the prostrate body up enough for the sheet to insulate him from the ground beneath.

Niall whips off his black neckerchief, provided to all firemen as a basic protection from the hot embers and hot water present at fires, and puts it to its other unofficial use, as a 'germ filter'. He places it over the man's mouth before beginning oral resuscitation, his tilted head observing the rise and fall of the unconscious man's chest as he forces his life-giving breath down his windpipe. A second ambulance is now on scene and its crew augment Niall's efforts, continuing them on the high-speed rush to hospital.

The ADO and our own Station Officer have been able to verify from neighbours that normally there are five occupants and, unless they had visitors at the time of the fire, there is only one person still unaccounted for.

Every inch of a fire building has to be searched. Even if no one has indicated that there are people still inside, a precautionary search is made. But where there are persons reported, the search assumes a greater criticality. It calls for experience, patience, skill and persistence if victims are not to be missed.

Sadly, the search does not take long. The conditions within the burned house, completely untenable only 30 minutes earlier, have now improved to the stage where it is possible to work without BA. The venting of the huge body of fire and heat, and the associated products of smoke and poisonous fumes, has created breathable, if not pleasant, conditions. The upper floors still feel like the inside of an oven and the familiar, pungent odour of a burned building remains.

Rising above the familiar smells of scorched plaster and brick, and charred timber now soaked with water, is a sickly sweet smell. However infrequently it enters the nostrils in a career fighting fire, it is an odour that is unforgettable: the smell of burned human flesh.

Even in the world's busiest fire departments, where fatalities and injury from fire occur regularly, it is certainly not every day that men witness death; in fact, it is quite possible to go a long period and not be on duty when such an incident occurs.

But whether death from fire has been experienced a little or a lot, you never really become accustomed to it. Its finality can stamp images on the mind that are not easily, if ever, forgotten.

And so it is, on this icy morning in the streets of inner London, that Ben Tuke's powerful box lamp sweeps its white beam across the fire-weakened floor of a rear bedroom on the top storey. Its light highlights a foot or so of charred debris, but that distinctive sweet smell is strongest near the foot of a sash window.

In the increased light of a tripod-mounted searchlight, a pile of debris about five feet in length, and gently steaming, is clearly illuminated. Yes, it is debris all right, but not that of the burned wood, incinerated contents and spalled plaster of the house; it is the remains of a human.

To the untrained eye, it would be difficult, if not impossible, save for the peculiar smell, to distinguish the two. But those with experience, born out of responding to hundreds of fires over time, can discern that what appears to be a baulk of charred roof joist is in fact the remains of a body.

As I mentioned earlier, the effect of severe fire and heat upon the structure of the human limbs is often such that the fists and arms are fixed in the pugilistic boxer's pose, a little like the foetal position of a baby in the womb. It is almost as if, at this terrible ending of life, the body returns to the position it once occupied inside its mother, before first emerging into the light of day and life. Here, now, it is the dark of night and sudden death.

Thankfully, we can be sure for the majority of such fatalities that the victim would have been rendered unconscious and asphyxiated by the smoke and fumes before being burned.

This sudden death shifts responsibility from the fire service to the police and coroner, but it is the brigade's task to ensure that the scene is undisturbed. None of those present can know at this stage whether the death is a result of arson, accident or indeed if a murder has taken place and the fire set in an attempt to destroy the evidence.

But now that all five occupants have been accounted for, and a fastidious search of all parts of the property and the piles of debris have not discovered any more victims, the ADO feels confident he can send the 'Stop' message to Control.

At just before 3 a.m., the following message is received in both Area and Brigade Controls from ADO Monsal:

> *Stop [no further reinforcing appliances or equipment needed] – House of four floors, approximately 35 feet by 55 feet. 80 per cent of ground floor, 90 per cent of first, second and third floors, and staircase from ground to top floor damaged by fire and smoke. Two women and one boy rescued by brigade from front room at second floor via Escape Ladder, suffering from the effects of smoke. One man rescued by BA wearers from rear room on top floor via internal staircase, overcome – all removed by ambulance. One man found in rear room on top floor apparently dead awaiting removal. Three jets, two hose reels, Breathing Apparatus. Same as all calls [all repeat 999 calls referred to the same fire].*

The ADO is now free to go and hands the reins back to our Station Officer, congratulating him on the good work carried out in the rescues.

As the Pump's crew, we still have some work to do, but the escape ladder is made up and returned to the station, it being the policy to place such lifesaving ladders back onto availability as soon as is practicable.

Fires of the sort we have just dealt with create a tremendous adrenalin rush, fuelling the energy and strength of all involved. Once the key actions of rescue and firefighting are finished, though, this adrenalin evaporates, especially in the small hours when it is our natural inclination to sleep. Actions then become far more of an effort, but the work required to ensure the fire is fully extinguished is of vital importance, as it would be a poor reflection upon the brigade to be called out later to find the fire had reignited.

We spend the next hour before being relieved by fresh crews from around the area cutting away any area of timber that might be hiding fire and heat, and turning over and damping down internal debris, save that in the area of the fatality.

One of the biggest hazards is posed by what we call 'bull's eyes'. These are round embers that look like red glowing eyes in the dark; they have to be cut out, as they could otherwise flare into flame at a later stage. Ceiling laths have to be pulled down using a long-handled pole known as a ceiling hook; plaster has to be stripped away to ensure that no hot sparks or embers lurk unseen behind it, and all the hot debris has to be thoroughly drenched and cooled.

It always has to be borne in mind by crews that a gallon of water weighs 10 lbs: a fire-weakened floor, already overloaded by a partially collapsed structure, could easily collapse with the weight of water used, plunging men through several floors. It is usual then, on the floors above the ground, to shovel charred debris either out of windows or into bins to be carried outside and then thoroughly soaked down. It is also good firemanship to replace large diameter hose with smaller hose lines, or to use only the rubber hose reel tubing, as soon as the situation permits, so as to reduce the weight of water used.

Whilst this is going on, Station Officer Tuke busies himself with the key task of trying to accurately pinpoint how the fire had started.

In due course we are relieved and finally return to the station at just after 4.30 a.m. The standby Pump, which has been mobilised to provide fire cover to the area denuded by our hours at this incident, remains until men have got cleaned up. A hot, smoking working job leaves you soaked in perspiration and your fire gear soaked in water from the firefighting process. Faces are streaked carbon black; hair is hot, matted and flattened by the helmet to the head. Eyes are red with the smoke and grime of fire smoke, and the mouth and nostrils are caked with dried saliva and mucus, the end product of eating smoke and hard physical graft. How good the hot shower and soap feels, and how pleasant to change from wet, sweaty T-shirt, trousers and socks into a fresh change of clothes – although the peculiar smell of fire smoke clings long to the skin.

We then replenish the equipment we have used, service and clean our BA sets, and then, only then, can we inform Control that we are back 'on the run' and allow the standby Pump to leave.

By 5.45 a.m., both Pumps are fully operational. Two or three of the watch who are particularly spent by their efforts take the option of resting on their bunks, but the remainder gather around the mess table for that great rejuvenator, the mug of hot, sweet tea and a post-incident discussion.

'That was a bit of a goer, fellows, wasn't it?' says Dec Mayo, the watch probationer.

'Nah, you haven't seen nothing yet, mate,' quips Geoff Joynt in a good-natured put-down, all part of the unwritten law, preventing inexperienced newcomers becoming too cocky – it's something akin to the armed forces' jibe pertaining to length of service: 'Get some in.'

'Actually, young Dec, it was going well, and I would wager if the fire wasn't deliberately started using some flammable material, there must have been paraffin heaters involved again. Every fire in a building I've attended that has spread as rapidly as this one was either as a result of accelerant use by arsonists or was in a factory in which flammable materials were present.'

'You're probably right, Geoff. The way that fire went up the staircase like a rocket and the degree of plaster spalled away, along with the depth of timber charring, does point to that,' chimes in Jim Peating.

'It certainly was hot in there, working the large diameter hose up the stairs. The way that the "stiff" was charred indicates great heat as well.'

'Andy and Jim did a great job with the rescue above all of that heat and flame,' adds Geoff, who had been instrumental in getting that vital water supply coupled up so rapidly.

'I'll be blushing in a minute. Won't we, Jim?' says Andy. 'But we all did a good job, and without you getting into the hydrant so fast, Geoff, to supply Niall on the covering jet, and the others entering from the front, our position would have been very iffy.'

Fire Investigation

Following all fires attended by the brigade, the officer in charge of the incident is required to complete a variety of reports and forms, and every fire involving a structure necessitates the filling in of a Home Office report; of particular importance in this document was the section headed 'Supposed Cause'.

Reporting the actual cause of a fire is useful for a number of reasons, but where there is a fatality, injury and persons are rescued, the importance of detailing the source of the blaze is magnified; it is magnified even further should there be suspicions that the fire was ignited maliciously.

In addition to some involvement at the Coroner's Inquest, which exists to determine the cause of a sudden death, and which normally would result in the officer in charge and the firemen who discovered the body giving evidence, a brigade representative could be called many months later to attend a hearing in a higher court, either in respect of criminal or civil law proceedings.

The whole outcome of such cases could rely very heavily indeed upon what the officer in charge had concluded to be the cause of the fire in question. In cases involving litigation against individuals, companies or the fire authority itself, the contents of the fire report, associated statements, photographs and the all-important 'real time' recorded radio telephone messages from incidents could be crucial to outcomes.

In addition to these 'judicial' and 'semi-judicial' reasons, the causes of fire also play an important part in the public educational process in respect of fire prevention and fire safety. If the source and cause of fires can be accurately pinpointed, then the resulting publicity can be all the more precise in its targeting of intended audiences.

Although fire brigade officers were aware of their responsibility to try to ascertain the cause, the heading 'Supposed Cause' was not always as helpful as another more precise form of heading might have been. 'Supposed' has ramifications of an educated guess; it encouraged a cause to be entered that, whilst possible, could perhaps be seen as not as probable as what at first might have appeared to be the case.

Where crews had spent several gruelling hours absorbing heat, swallowing thick smoke and becoming soaked to the skin, it was not too hard to adopt only a cursory approach to the 'fire cause' investigative process. Firemen could be extremely persuasive in their requests to get back to a station for a shower and change into dry clothes. At such times, a Station Officer had to have a strong personality and adopt a professional approach. If not, he would end up looking for evidence to try to support a particular theory as to how the fire had started rather than remaining objective.

It was possible also to enter 'Unknown', and with a rising number of arson incidents being experienced nationally, the previous somewhat imprecise process of investigation began to be replaced by a much more forensic approach, especially in the busier metropolitan fire authorities. Such was the development by the fire services, working in conjunction with the police and scientific forensic services, that eventually 'Unknown' would be quite a rarity on the Home Office fire report in relation to the cause.

Not only would this increased accuracy in determining the causes of fires have beneficial effects upon fire safety drives, but it also assisted in the detection of arsonists, lessened the potential for insurance companies to be defrauded and helped bring perpetrators to justice.

Within London Fire Brigade, and an increasing number of provincial brigades, specialised and dedicated Fire Investigation Units were introduced and staffed by experienced officers, who possessed an aptitude and interest in this branch of work. They brought all of their previous experience in firefighting and rescue, and applied that knowledge. This difficult task was aided by the use of a range of equipment such as highly sensitive accelerant detectors, even 'sniffer' dogs, to assist the officer in charge to arrive at an accurate assessment of how a fire had begun and of how it had spread.

In attempting to find out what had caused the fatal fire that we had been called to on that freezing cold morning, Station Officer Tuke didn't have the luxury of a sensitive device to 'sniff' the scent of petrol, paraffin or other accelerants. Accordingly, and until the forensic investigators from the police arrived, he had to rely upon his own nose and eyes, drawing on his considerable experience of previous fires and of the telltale signs to establish not only where the source or sources of the fire might have been but also whether it had developed slowly or quickly.

In the former, he will have noticed whether any window glazing or internal mirrors, not shattered by the fire or by firefighting operations, are heavily sooted or smoke-stained. This indicates a slow, smouldering fire that has produced heavy smoke over a long period. In the latter, he will be looking for an absence of smoke-staining and for clean breaks of glass, which can point to a rapid build-up of heat, as is the case if accelerants have been used.

Flammable liquids, poured onto floors and stairways, even through letterboxes, produce on ignition a rapid fire with a high temperature of flame and superheated smoke. These high temperatures lead to the deep charring of wood and can cause wall plaster to fall away. Also, in the area of initial ignition, they produce a definite lightening of exposed brickwork, as well as a pattern of burning.

Arsonists often start fires in more than one part of a building in their quest to ensure the maximum damage and spread, and the fire with more than one seat is one of the clearest indications of deliberate ignition.

All of these developed skills and empirical fire-ground experiences are in the forefront of the Station Officer's mind as he sets about the tedious process of investigation. Conscientious by instinct and developed practice, he applies as much diligence and tenacity in attempting to locate the actual cause of fire as a keen police sleuth applies in finding out who has committed a murder.

He had made a mental note immediately on our arrival several hours earlier of the rapidity with which the fire was developing, as well as the colour of the smoke, the speed and force with which flames and smoke were issuing from windows at nearly every floor.

Later, when providing reassurance to Jim Peating and me working the jet up the stairs, he had felt the great heat, and the springiness of the floors and stairs, pointing to structural weakening caused by very hot flames.

Now, as he checks the blackened, steaming interior from top to bottom, he can see little smoke-staining on dressing table and wall mirrors, but notices how much plaster has come away, particularly in the area of the ground-floor hallway and along the whole walls of the staircase from top to bottom.

In his notepad, he records that an electric wall clock in the kitchen has stopped at 1.35 a.m., roughly the time of our arrival. He notes that, in the understairs cupboard, all of the fuses to the electrical supply are intact and of the correct amperage rating for power and lighting circuits respectively.

He had spoken to the immediate next-door neighbours earlier and they had confirmed that paraffin heaters were used in the house – and it would have been especially true at that time, as Britain felt the freezing blast of a cold front from Siberia.

In the hundreds of house fires that Ben Tuke has attended over the years, many of them involving the paraffin stove, he has seen far more times than he can remember fires started by heaters placed either at the bottom of the stairs or in hallways and landings. He has even been to several fires in which the occupants had placed the smaller stoves under their beds!

He therefore concentrates his efforts on the ground-floor hallway, about six feet away from the start of the dog-legged staircase, an area in which he has already observed much damage to plaster, deep charring of the staircase treads and risers, and a section of exposed brick on the right-hand wall, which was the party wall with the neighbouring house. Rising from the blackened and still steaming ten-inch-deep pile of debris adjacent, there is a definite outline of a jagged shape. This mirrors what has been a flame – hot enough to cause the plaster to come away, and which had turned the redbrick to a whitish salmon-pink.

Station Officer Tuke kneels in front of this debris, his burly form hunched over as he gently probes the charred pile, a searchlight illuminating the scene. The small spade he is using so gently touches something metallic about five inches below the black detritus of the fire. He stops his gentle probe, puts down the spade and feels around. Uncovering more of the debris, he sees a thin wire handle, then the mica plastic window of what is clearly a paraffin stove.

The brigade photographer has finished his recording of the body and fire scene, as required by the Station Officer, some 20 minutes earlier, and is soon taking further shots of the heater and immediate surroundings. Three further paraffin heaters are found, one on the first-floor landing, another on the second-floor landing, and the third outside the room in which the fatality has occurred on the top floor.

Later, Ben Tuke confirms from those fortunate to have been rescued that the owner of the house lived in the flat at the very top of the property and had regularly placed paraffin heaters in the locations in which they have been discovered. He also establishes that the owner had two cats, which roamed freely about the place.

In due course, he completes the 'Supposed Cause' section of the Home Office report in respect of this tragic blaze:

> Filled and ignited paraffin heater placed in hallway near foot of stairs knocked over, probably by domestic cats, causing contents to spill and ignite stair carpeting and surroundings, including contents of three further paraffin heaters on first, second and third floor landings.

A month later, Station Officer Tuke and the BA wearers who had discovered the victim attended the St Pancras Coroner's Court

Inquest, where a verdict of accidental death was recorded.

The two cats were never seen again. They probably escaped by leaping from an open window and roamed about as wild cats, or were taken in by someone somewhere, with whatever remained of their 'nine lives'. Their keeper, the landlord, in the small top-floor bedsit had only one life. That ended on an icy morning in north London, yet another victim of the far too many fires caused by paraffin heaters during this period.

Chapter 21

A Night to Remember: Docklands, 1973

Fire in the London Docks

Old Father Thames is flowing by the mouth of Heron Quay,
cats and rats are fighting near the lock gate to the River Lea.
Two tall beat coppers are patrolling, in the alleyways besides the wharves,
where massive powder keg-filled warehouses
make them both appear to be dwarves.
Suddenly from out of the still night comes the sound of a shattering pane.
It echoes from a long black alley that is known as 'Whoring Lane'.
Their plod now turns to a gallop, towards where the crashing came,
and as they enter the alley they meet a massive wall of flame.
Now's not the time to dawdle, there's a warehouse burning a treat
and their radio crackles with a message which echoes in the deserted street.
It's an urgent, breathless transmit to report that a big fire is about,
and this rapidly leads Police Control to give the Fire Brigade a shout.

Call bells are doing their calling across the London in the East.
Powerful diesel engines are revving, sounding like a roaring beast.

> Bleary-eyed men are rigging in helmet, tunic and boots
> as eight tons of Pump and Ladder out of the station shoots.
> The Guvnor's hand is grasping a slip of paper coloured white,
> On it the cryptic message tells of a docklands inferno raging in the night.
>
> – Anon

The London Docks, once the pre-eminent port on the planet, were in terminal decline when I became a London fireman in the early 1970s. The wharves, ships, boat-repair sheds and warehouses that fell within the boundaries of the London Fire Brigade had posed many challenges for fire crews over their long and illustrious history, which ranged from the conflagrations during the Blitz of the 1940s to the serious peacetime fires that occurred in the massive warehouses jam-packed with all manner of combustibles.

The London docklands had suffered some of the heaviest damage of any British city during the bombing raids of the Second World War. The firemen of that period had, to quote former Chief Officer Cyril Demarne OBE of the West Ham Brigade, 'packed a lifetime's experience into a matter of months'. Then, whole rows of warehouses and riverside properties were alight; surrounding streets had been rendered impassable by melting paint and tar – even the most powerful jets of water were largely ineffectual against the massive fires that developed.

Given the huge historic importance of London's docklands to the economy of the capital and to the nation, I never responded to emergencies there without feeling a great sense of what had gone before.

From the early 1970s right up to the late 1980s, I had a close association with these atmospheric districts – I only had to scan the pages of my LFB-issue blue-bound street atlas to visualise the countless responses made to addresses within and around the docklands; I could easily bring to mind the harsh living conditions I'd witnessed that so many hard-working East Enders had endured.

In later years, when involved with the innovative and towering developments brought about by the Margaret Thatcher-inspired regeneration of the docklands, it was hard to believe how things had changed since my arrival in the early 1970s. I often wondered what those who had lived and worked on the Isle of Dogs would have thought, had they seen the results of such burgeoning progress. What would the tough, spirited housewives shopping at Chrisp Street Market in Poplar have said?

Attending fires and non-fire incidents in warehouses, small lock-up factories, crumbling tenements and garment and furniture factories within that big U-shaped Thames loop, from Millwall in the south up through Manchester Road to the East India Dock Road onto Limehouse, Ratcliffe Highway, the once prostitution-scarred Cable Street, then into Whitechapel and the Mile End Road, provided an insight into the docklands and its surrounding districts.

I felt a sadness, as the fires we fought within those warehouses and wharves became a flaming reminder of a once-proud industry in decline. Indeed, as the flames consumed roofs and floors, and walls crashed to the ground amidst smoke and steam, we seemed to be seeing the docks in their final death throes – never a pretty sight.

* * *

The heavy steel sliding fire door rolled along the sloping rail on which it was hung and slammed shut with a metallic crashing thud.

'This door will hold back fire and smoke for as long as the thick wall in which it is set,' booms the strong voice of Station Officer Jim Cronine to the four assembled members of the Pump's crew, who were taking part in an outside duty known as a 'One One D' (it was sanctioned by Section 1(i) (d) of the 1947 Fire Services Act).

Such visits were organised in order to familiarise crews with selected premises. It would have been a practical impossibility to visit every building on a station's ground, so we concentrated on those that held the greatest potential for a serious fire, were of a large or unusual construction or had a high life risk. Falling into these categories were the major hospitals, factories and department stores, and also prominent buildings such as St Paul's Cathedral, the Bank of England, the Tower of London and the massive Thames-side warehouses, inside one of which the five of us were gathered at that moment.

The 'One One D' enabled fire crews to gain information on such things as the water supply, the presence of sprinkler and fire alarm installations, the materials stored or industrial processes carried out within the premises, the number of employees and their locations, and such internal provisions as fire-resisting walls, floors and steel fire-stopping doors.

These familiarisation visits were often combined within the inner London suburbs with inspections carried out under the London Building Acts (LBAs), which were regulations unique to the capital and which could historically be linked back to the Great Fire of London of 1666, when the dangers of small fires spreading from their source to involve neighbouring premises, then adjacent streets and thoroughfares, had been so vividly and catastrophically demonstrated.

A series of progressive fire-protective structural developments over the following 300 years or so, honed to greater effectiveness by conflagrations experienced within the capital, had culminated in the introduction of the London Building Acts. The 1939 amendments to this Act embraced any commercial premises whose cubic capacity exceeded preordained limits. It required the provision of automatic sprinklers and fire-resisting walls, floors, doors and shutters, the purpose of which was to contain the spread of fire and smoke long enough for the LFB to gain access and extinguish any fire occurring. The proliferation of massive warehouses on the banks of the Thames, along with other huge structures, often covering a very large number of floors, made these regulations vital to the process of fire protection in these high fire-risk localities.

The steel fire door that had slammed shut so loudly was one of four positioned on each of the six floors of this massive warehouse, only a river's width from the majestic structure of the much-loved and much-visited Tower of London and next to the Pool of London, as it used to be known. Then, cargo from across the globe found its way to the heavily fortified dockside complexes. These were huge, imposing buildings, designed to protect such goods as ivory, silk, spices, timber and precious metals from the pilfering and wholesale theft that were so rife before their construction.

The LBA requirements were such that any fire starting would be held in check by the actuation of the sprinklers and by non-perforate walls. These, along with the steel fire doors and shutters, would prevent any fire and smoke spread outside the compartment of origin. Each door ran along a slightly sloping rail. It was held in the open position by a counterweighted cable. At the centre of the cable there was a fusible link: a soft piece of metal designed to part on exposure to the heat of a fire, causing the door to be freed from the counterweight and slide down its rail, thereby closing the opening in the thick fire-resisting wall.

'Of course, arsonists have been known to wedge these fire doors in advance of their horrible deeds, so as to prevent them from doing their job, in addition to shutting down the water supply to the sprinkler system,' Jim Cronine goes on.

'So they torch the place and the fire spreads throughout the warehouse?' queries Sevvy Malt, a former policeman. In only his first few months out of probation, he was proving himself to be a keen and capable addition to the watch.

Jim had been referring to the spate of fires that had occurred recently in some of the warehouses in this locality. The area was at the time prime real estate and many businessmen were anxious to lay their hands on the highly sought-after Thames-side land; there had been rumours circulating that some of these fires had been of a 'suspicious' origin, which is police and fire brigade speak for arson.

Only a few months back, we had attended a ten-pump blaze nearby. It had been a rapidly spreading fire, which often meant accelerants had been used to hasten the spread. It had also occurred on a Saturday evening, when the chances of its detection would have been less than in the week. I had been a designated BA wearer on that occasion and remembered the incident clearly, because as we waited to receive the order to start up our sets and make our way via an escape ladder to the second floor, a flashover had occurred inside, catching a crew who were working a jet there.

Not the sort of thing to build up confidence.

Within seconds, a ladder had been re-pitched to the affected level and in minutes the Sub Officer in charge of the crew was being carried down. Fortunately, his burns turned out to be only superficial, but after the flashover the officer in charge had made the decision to withdraw all crews and we spent the next few hours fighting the blaze from the street.

'You've got it, young man,' Jim continues. 'And don't forget, if the doors to the stairs are wedged open as well, the fire can spread from floor to floor, a real nightmare scenario, yeah?'

Jim Cronine had been to many nightmare scenarios during his long years of service, the vast majority having been spent in and around the East End. Like Ben Tuke, Biff Sands and many of his peers, he had seen it all and nothing seemed to faze him. Although not a man you'd want to cross or question when out on the fire ground, he was a considerate individual who cared about his men.

We rounded off the tour and headed back to the station.

None of us present on that outside duty that morning would have envisaged that we would one day be returning to this self-same structure to a blaze that the television newscaster would describe as 'one of the largest fires to be experienced in London since the Blitz'.

It is a long-standing custom within the London Fire Brigade, especially during the 15-hour stretch of night duty, to have a mug of tea following the six o'clock roll call and the routine check of lockers and breathing apparatus. Unless we are called out, the evening meal is always scheduled for eight o'clock. The tea is a little hot liquid nourishment prior to commencing drills, technical lectures or going on outside duties, such as hydrant inspections or street familiarisation.

Therefore before running up the stairs to the mess room for my tea, I have carefully checked the breathing apparatus set to which I have been assigned, along with Sevvy Malt and John Lunsett. Together with Jim Breen, the driver, and Station Officer Jim Cronine, we make up the Pump's crew.

Even though the preceding shift may not have involved the use of BA, it is still the individual wearer's responsibility to carry out his routine checks, including noting the amount of oxygen remaining in its cylinder and the satisfactory operation of a warning whistle automatically sounded to remind the wearer that the oxygen supply is nearing exhaustion, and basically giving it a general once-over.

That first gulp of the hot, tannin-brown, sugar-sweet Rosie Lee has barely coated my throat when the urgent ring of the call bells reverberates across the mess room. Within 30 seconds, both appliances are thundering through the tall canyons of the City of London's streets. I notice as we hurtle forwards that the slipstream of the two engines had thrown up a discarded newspaper into the air. As we near Tower Hill, the unmistakeable pungency of fire smoke stings our nostrils.

'Probably some bonfire set alight before tomorrow,' shouts Jim Breen, tugging at the gear lever. It is the night before 5 November.

Jim's surmise might have been correct had we been responding to a call-out in the residential suburbs, but that's not where we're heading. 'Bonfire round these parts, Jim? Don't be a plank!' retorts Sevvy Malt.

'Christ, if it's a bonfire, it's a bloody big 'un, mate,' chips in John Lunsett. 'Look at the effing smoke!'

A heavy, swirling fog, looking a mustardy-yellow-brown in the dimmed-down reflection of the tall street lamps, is banking to pavement level. In the 1950s and 1960s, London was notorious for its pea soup smogs, which blanketed buildings, people and vehicles in its suffocating veil; this isn't then, and this isn't a London fog in that sense. It is a fog, though. A hot, pungent, rolling brown blanket whose texture is so closely woven that both Pumps are forced to brake, so reduced is the visibility.

And then, a breeze of night air, funnelled into a gust by its passage through a narrow alley, draws back the smoke curtain to reveal centre stage: with the majestic Tower of London in the wings, it is the self-same Thames-side warehouse we had visited only months before. With Jim Cronine's words about arsonists returning to our minds, we see ominous huge inverted cones of thick, dark smoke racing out of windows on all visible sides and shooting rapidly towards the dark November sky.

The two appliances haven't yet come to a halt when the radio's metallic voice sings out, '*Priority from Station Officer Sands – Make Pumps Ten, Turntable Ladders Two.*'

'Fuck me, a straight ten – we've got a real goer here, boys,' shouts Sevvy Malt, whose enthusiasm for 'make-up' fires is a facet of the ex-policeman's keenness. This enthusiasm is not born out of any wish for anyone to suffer; rather it is the same sort of enthusiasm displayed by the military when they are being dispatched to a trouble spot: there they will be able to display at long last all of the skills learned via hours of simulated exercises and drill manoeuvres, and feel the adrenalin rush that springs from raw, real-life scenarios, rather than the oft stale and repetitive practice runs.

At many fires, the initial officer in charge, conscious of the numbers of appliances on the first attendance, will be conservative in his requests for extra appliances. Often the real extent of a fire is not evident on arrival and a quick survey has to be done to avoid an embarrassing situation as a result of a premature 'make-up' message. I remember one call in the City of London where a Station Officer made pumps eight before checking out the source of the flames – it turned out to be a weather protection covering on scaffolding alight, which was soon extinguished with a hose reel. The Divisional Commander was not amused!

A straight make-up request calling for seven more pumps plus an additional Turntable Ladder is a sure indication that a potential conflagration is afoot. Especially when a man as seasoned as Biff Sands is the one who initiates it.

Like so many of his era, he does not initiate a make-up message lightly. He is of a breed that has such pride and confidence in the competence of his men that every effort will be made to 'stop' fires without needing to call for reinforcements. In fact, it is almost a form of management psychology: by convincing his crew that they are the most efficient firemen around, they believe this and live up to it.

It is also an almost jealous guarding of his patch. 'Make-up' fires always result in a supervisory officer coming on from Divisional HQ, a few of whom do not measure up to Biff Sands' strict criteria of being a 'good hand'. As a consequence, 'make-up' messages are only sent after his developed instincts have left him in no doubt that without additional pumps and men the fire will 'get away'.

That Biff was correct in his initial assessment is becoming evident with each passing minute. A glance at all of the warehouse windows over the whole frontage of the five-storey warehouse tells its own story, as the squawk of the radio confirms Biff Sands' informative message: '*From Station Officer Sands – general warehouse, five floors, approximately 200 feet by 200 feet, 75 per cent of all floors alight.*'

From each of the square prison-like windows, inverted cones of brown and yellow smoke are funnelling upwards at a speed that suggests only one thing: arson. Just as our own guvnor had intimated in February, during our familiarisation visit. No matter how combustible the contents, fire would never be able to develop so rapidly, especially in a building like this, which is subject to the massive fire-protection provisions of the London Building Acts. In fact, it now seems extremely likely that someone must have entered the warehouse over the weekend and set in motion exactly the sequence of events the guvnor had warned about.

Station Officer Sands' men have already got two large diameter jets to work, their silver streams playing into the windows in some effort to quell this fiery monster. Our own Station Officer is liaising with him and, as we await instructions, we look up at the belching smoke and see a ruddy glow moving in a veiled dance behind the dark curtain of smoke. Barely visible, and initially only at the first floor, the glow, like that of an electric stove ring on its lowest setting, gradually begins to appear at all floor levels. The redness then starts to further increase, looking like the stove ring now turned up to its maximum temperature.

Then, as if liberated from its smoky shackles, the vicious fire bursts forth, turning the darkness of the November night into almost the light of day, casting a vivid, shimmering reflection on the adjacent cold, black water of the river basin to the east.

'There goes the eggs, beans and chip supper,' quips Sevvy Malt, secretly bubbling inside at having 'picked up' a blaze of this magnitude.

'You're not kidding, mate. And you can probably say cheerio to breakfast as well, looking at this bastard,' says John Lunsett, who has himself attended some of London's largest infernos in his 17 years of service.

The duty ADO from Divisional HQ arrives, his white car screeching to a stop. He alights hurriedly, flinging open the cavernous boot from where he lifts out his fire kit, into which he rigs, balancing precariously on one leg as he moves from shoe to boot.

He is a short, youngish man, probably a decade younger than Biff Sands, towards whom he purposefully strides to get an appraisal – not that the scale of the fire isn't plainly visible to all. The duty ADO knows that a make-up of ten pumps, plus the additional Turntable Ladder, will see the duty Divisional Officer (DO) from Shoreditch on scene and, in view of the rapidly deteriorating conditions, knows that a further make-up message calling for an additional five pumps won't be too many.

The ADO's '*Make Pumps 15*' is received just as the grizzled form of the duty DO arrives in his chauffeur-driven car. He finishes off the cigarette that is seldom out of his lips, quickly stubbing it out in the car's ashtray, before clambering out, temporarily placing both of his huge hands across the small of his back as he does so.

We have been instructed to lay out four large jets. With these, we are to mount a 'from the street' onslaught into the burning warehouse in an attempt to lessen the searing waves of heat radiating from it, which are quite capable of igniting the office block adjacent, even though it is a good 30 yards away. The terrific heat being generated by the rapid burning of such a huge fire is sending clouds of smoke and huge chunks of white-hot firebrands and embers hundreds of feet into the night sky. This is creating real concerns that some of these embers might set fire to one of the capital's prized and historic monuments, the Tower of London being only a few hundred feet away.

'Effing hell! If the Tower lights up, the Chief Officer will end up in there,' jokes John Lunsett.

'Where will that put us, then?' chuckles Sevvy Malt. 'We'll be bloody burnt at the stake!' he adds, struggling to control the jet, with me leaning onto the hard canvas of the hose. The jet is working at over 85 lbs to the square inch pressure in order to throw the solid white lance of water into the shimmering maw.

The fears for the Tower and the many other commercial premises in the vicinity are such that two of the reinforcing crews are detailed to slowly tour the surrounding streets, to look out for fires started by the showers of embers, whilst the security staff and guards within the Tower are alerted to carry out a similar task.

The DO, ADO and Station Officer Sands have by now done a recce of all accessible sides of the warehouse. It has become clear that the fire involves a good half of the building at all levels, but it's possible that the fire doors might be still in the closed position in the rear portion. This suggests that there is the potential to make a bridgehead against the blaze at the landing levels at the head of each staircase. There, BA crews can get large jets to work at these positions as soon as possible. Given the rapidly developing conflagration and the manpower that will be needed to provide the water and emergency standby BA teams, the DO sends a further message, making the pumps up to 25.

At all fires requiring 15 pumps or more, the policy is for a Principal Officer to be ordered on from Brigade Headquarters. The Assistant Chief Fire Officer (ACFO) is telephoned in his penthouse flat, which overlooks the Thames at Lambeth Bridge on the Albert Embankment. He looks out towards the illuminated face of Big Ben. It's ten minutes past seven. Immediately taking the lift down to the appliance room, where his staff driver is already revving the engine, he is on his way, anxious like all firemen to get to a major blaze.

As the Assistant Chief is driven across Westminster Bridge and then turns right onto the Victoria Embankment, he can see the huge, red shimmering light illuminating the eastern sky. It must look for all the world like some of the scenes he had witnessed over 30-odd years ago as a young fireman involved in fighting the tremendous blazes caused by the Blitz.

There are more than twelve large jets playing into the windows now from the ground, plus the water from the monitors atop the two Turntable Ladders. These are sending a crashing deluge into the upper floors. The monitors deliver hundreds of gallons a minute and their capacity for absorbing heat is tremendous. But there is a downside. Not only does the firefighter's main ammunition weigh over 10 lbs a gallon, but in these old warehouses the floors are supported by cast-iron columns and, as firemen know, cast iron has tremendous strength in compression but it has a fatal weakness in that its tensile strength is low: the sudden cooling of a cast-iron column can cause it to crack and collapse, leaving floors laden with heavy stocks of goods partially unsupported. It only needs the additional weight of thousands of gallons of water to cause a sudden and massive structural collapse. It is for this reason that whenever fire crews have to enter a burning building for firefighting or rescue purposes, jets are usually shut down.

Our guvnor strides over after visiting the Control Unit. 'Right, listen in. BA men are to get rigged. Five teams are going in via the south-west corner entrance. One team is going to each floor, but only in sequence. That is, no one goes to a floor above until the team below have got their jet onto the fire. At the bridgehead, any closed fire doors are to be carefully opened, just enough to get a jet to work. If the fire is too fierce, close the fire door and retreat – any questions?'

An identical briefing has been given by the Station Officers of the other four teams, and after others have taken over our jets we go over to the Pump Escape and Pump to begin the tedious process of donning and starting up our old friend 'Proto'.

As I go through the start-up procedure, taking the small personal head harness from under the inner shell of the helmet that secures the mouthpiece to the face and head, and go over the drill book ditty – 'while the dresser buckles you . . .'* – I look up at the massive inferno confronting us. I glance across at Sevvy Malt and John Lunsett. The brown rubber goggles pushed up onto their foreheads reveal a sense of apprehension in their eyes, an

* This is part of a rhyming memory aid printed in the official Home Office drill book to assist BA wearers in carrying out the correct donning and start-up procedure.

apprehension I can feel in the pit of the stomach. No matter how well trained a fireman is, no matter how experienced, one can never get away from the fact that we are heading into the unknown. Especially this evening, as we attend a 'once in a decade' conflagration. We are placing our trust in 30 lbs of breathing apparatus to safely see us through the heavy rolls of superheated smoke, now almost viscous in its appearance, and with the searing heat of fire all around us.

The BA entry point is set up and we hand in our tallies, then prepare to begin the hard graft of working the heavy charged hose lines up to each floor, as instructed by our Station Officer. No sooner have we taken the first steps towards the south-west entry when a ripping, tearing crash rents the smoky atmosphere aside. A huge portion of the top three floors of the warehouse cracks open, creating a huge split about a foot wide, narrowing to a few inches 30 feet below. Angry tongues of white-hot flame blast out as if from a gigantic blow lamp. Our hearts are in our mouths, especially those positioned with their jets at the northern face of the warehouse, where the huge fissure has occurred. We all freeze, fearing the whole facade will collapse onto us.

'My God! Not another Cheapside,' I think, as into my mind comes the thought of a terrible fire in a Glasgow whisky bond in 1960 that took the lives of fourteen firemen and five locals from the Glasgow Salvage Corps, when the massive stone walls of a similar structure collapsed onto them.

'All crews withdraw! All crews withdraw!' shouts the urgent voice of the duty ADO through a loudhailer.

The ACFO from Lambeth has only just finished being appraised of the situation by the DO and has just assumed command when the huge crack appears. Although the fireman's primary duties are to save life and property, there is a time when the safety of personnel becomes paramount. Especially when there is no doubt whatsoever that there are no occupants reported to be inside a building and when such a massive structural collapse has taken place.

The Assistant Chief doesn't have to think twice about giving the decision for all crews to be withdrawn. His instant command decision arose not only from his expertise, developed over years, but also out of a genuine concern for the safety of crews now under his command. He has been present at too many harrowing visits to inform relatives of firemen killed at incidents and has attended too many of the Brigade funerals that followed to risk his men in this conflagration.

Above all, he knows that this is a massive fire and one that is not much different in size from those of the 1940s, when it would have been impossible to do anything but fight the blaze from the building's exterior.

The urgent order for withdrawal obviously means that our own task of gaining entry to attempt to cut off the spread of fire is immediately aborted.

'Saved by the bleedin' bell there, all right,' says Sevvy Malt, his speech no longer impeded by the mouthpiece, as we close down our BA sets and unrig.

'Not half, mate. Another few minutes and we'd have been inside. It don't bear thinking about. Although it ain't collapsed yet,' says John Lunsett.

'No, but I bet it won't be long before it does,' Sevvy goes on, as we carry our sets back to the Pump, now surrounded by at least 15 others and amidst a web of snaking hose lines.

Such is the concern of the officer in charge that the front portion of the warehouse might suddenly collapse that he has detailed officers to begin a retreat of those appliances parked closest to the weakened structure. This is no simple matter. The first couple of Pumps to arrive had set into the street hydrants straight away in their urgent need to get jets onto the blaze. The whole area of road and pavement behind them is a mass of charged hoses supplying the jets to quell this conflagration. But there is no option, unless we want to see several front-line appliances crushed under tons of debris. The hydrants feeding these are shut down, along with the firefighting jets, then the disconnected hose dragged away to allow the Pumps to be reversed out of the path of the collapse, which is looking more imminent each minute, as huge chunks of masonry crash to the ground. Of course, the premature shutting down of these jets gives this massive blaze the chance to grow even bigger.

It takes about ten minutes of frantic effort before the Pumps can be reversed and it isn't a moment too soon. Very shortly afterwards, a massive section of wall collapses and floors from the top down to the first break off from the gaping canyon of a crack and hit the ground with a huge 'crump', throwing up a huge dust cloud to mingle with the smoke.

Virtually everything that Jim Cronine had talked about on that earlier visit to this warehouse appears to have happened. The sprinklers must have been shut off, and all of the fire doors wedged for a fire to have spread to this size so quickly. Petrol has probably been used: the speed with which those inverted cones of smoke had been shooting out of most windows, along with the ruddy glow on all floors on our arrival, pointed clearly to the use of such an accelerant. The intense heat produced by the use of petrol will have quickly ignited the tons of combustibles on each floor. In the absence of sprinklers, and with the fire doors wedged open, the super-heated gases and smoke would soon have raised the internal temperatures to a level where everything combustible bursts into flame. Such terrific heat expands steel floor joists and burns through timber floors, the combined effects leading to structural collapse, as walls are pushed out by the forces created from weakened floors loaded with heavy amounts of goods.

Now that any hope of cutting off the internal firespread has gone, all our efforts have to be concentrated on getting the fire surrounded, applying such a force and amount of water that the risk to surrounding properties is reduced as soon as possible. Accordingly, the Assistant Chief requests a further ten pumps, making this blaze one of the biggest attended since those long nights of the 1940s when 'all of London seemed to be alight'.

Opposite the northern collapsed wall of the warehouse is an office block and its flat-railed-edge roof provides an ideal position from which firefighting jets can be positioned and directed. The 100-foot-long General Purpose (GP) lines are taken up onto the flat roof via the internal staircase, the keyholders having been alerted much earlier. They are anxiously praying that we will save their office block from the radiated heat, flying firebrands and embers falling from the sky.

The GP lines are lowered down to the road and the uncharged

large diameter hose and branch pipes are hauled up to the roof. Before the 'water on' instruction is given to the Pump operators far below, each of the half-dozen or so jets is lashed to the stout steel tubes of the railings that guard the edge of the roof, making the job of directing the powerful jets far less strenuous – the weight of charged two-and-three-quarter-inch-diameter hose makes for heavy work, especially when water is being pumped at high pressure in order to overcome the frictional resistance in the water ascending from the street, as well as being thrown the 60 or so feet onto the blazing warehouse opposite.

'That's a bloody good jet, fellas,' says Jim Cronine. He's standing behind me as I direct it in a sweeping arc by moving the large nozzle in its lashing. It is of little use to keep the jet stationary: the idea is to always keep it moving slowly, thus knocking down flames and cooling down the heat over as wide an area as the lashing allows.

The fire is still of terrific proportions a good two hours after our arrival. But it only feels as if we have been there a half hour or so, such is the effect of adrenalin at incidents of this magnitude.

But as the urgency of our actions is reduced, so too are the levels of adrenalin pumping through our systems; at this point in an emergency, firemen have to dig into their physical and mental reserves, especially when fires are being fought from the street and it is becoming colder as the night slowly rolls onwards. Details that have gone unnoticed during the frenetic, heart-pounding activity of the first hours become more and more evident and annoying: a barely noticed leak of water from a defective coupling washer onto the hands, wrists and arms can seem like torture as the temperature falls over time; the same water and the spray-filled air that soaked the rear of our blue overall legs and which was not even felt during the first hour slowly becomes chillingly evident, soaking us through with a cold that penetrates through to the bones. The soaking is caused by the smart-looking, black polished leggings – smart-looking but extremely ineffective at keeping the back of the legs dry!

If there is one antidote to this somewhat miserable and wet state of affairs, it is the arrival of the canteen vans from Brigade

Headquarters and the Salvation Army respectively. Suitably refreshed, we return to our rooftop post.

Some 25-plus jets and monitors are now attacking the blaze from all sides and from aerial appliances sited at strategic positions, and in due course the officer in charge sends the standard message: '*Fire Surrounded*'. I glance at my luminous wristwatch and notice that it is just before eleven o'clock. We have been here from just after 6.30 p.m. As we were part of the initial attendance, we are due a relief soon.

Those Pump Escapes that could be relieved have already gone, including our own. They are still the Brigade's principal lifesaving appliance on account of the wheeled escape and it is Brigade policy to always get these back to their station, ready for other emergencies.

It is the officer in charge who makes the order for relief crews and appliances, and our wait will depend on a number of factors. If there are other serious incidents going on across the capital in which large numbers of Pumps are attending, then it could be a long wait since fire cover cannot be denuded over large tracts of London.

At a fire of this magnitude, it will be many hours before the 'Stop' message can be transmitted. The potential dangers to adjacent properties and to fire crews present from further structural collapses means that a proportion of the relief Pumps will need to have a Station Officer in charge for reasons of command, control and personnel safety. Pumps satisfying this rank requirement might have to travel across London from as far away as Feltham or Heston in the far west near Heathrow Airport and this can take time.

Crews will need to be kept at this fire for some days, their numbers gradually being reduced as the operations change from the quelling of the still huge body of fire right through to the damping down and cutting-away stages, with the District Building Surveyor on hand to advise which parts of the structure are unsafe.

Our relief arrives at just after midnight. So, some five and a half hours after we had driven into that hot brown fog of smoke, we are able to leave the fire ground.

The cold night air is still heavy with that unique odour of

burning, but the massive volumes of water delivered from all vantage points have finally begun to tame this conflagration and the dense smoke of earlier is no more. The ten o'clock news has reported the blaze as one of the biggest to have been seen in London since the Blitz. As I look over to the scene as the Pump heads back along Tower Hill, the comparison with that era, in which the London docklands took such punishment, is not entirely inappropriate. The massive warehouse, which we had known at its best, now looks like so many of those dockside buildings captured by wartime photographers: a once-proud structure reduced to a shadow of its earlier self.

But tonight, unlike then, no matter how hard we worked to suppress the inferno, no matter how palpable had been our apprehension as we prepared to take our BA-festooned forms inside the blazing building, there was no threat from the terror that rained from the sky down on our forebears. We could not have begun to imagine the feelings and fears of those nights, where the dangers of firefighting were magnified so much by the bombing raids, which could, and often did, obliterate so many, without regard for rank, age or gender.

It was now 5 November: 'Remember, remember the fifth of November, gunpowder, treason and plot . . .' The familiar nursery rhyme comes into my mind as the Pump slowly winds its way through the city streets, the acrid smoke still heavy over almost a mile from the blaze.

In another 18 or so hours, the Brigade will begin the annual ritual of responding to bonfires that have grown too large for a neighbour's comfort. We have seen enough flame tonight to ignite a million bonfires. The damage caused is immense, but the surrounding properties, especially the historic jewel of the Tower of London, have remained unscathed through the combined efforts of more than 200 personnel and 40 appliances. It certainly has been a night to remember.

Chapter 22

Them and Us and Things in Between

Such was the range of emergencies during the time I served in the busiest sharp-end stations of central London that it was possible, with a bit of luck – you had to be on duty when such jobs came in – that a fireman such as myself, with aspirations for promotion, could gain the solid practical experience required quite quickly. As explained, a deep and wide operational background was considered essential within the UK Fire Service, and the London Fire Brigade in particular, and was a primary motivation for my transfer from Yorkshire.

The fire service is strictly organised in terms of hierarchy and so promotion plays an important part in the career of a fireman – it also allows a culture of 'them' and 'us' to exist. Soon after my transfer south, I noticed a real difference from Yorkshire in terms of attitude and respect offered to the man who has taken on the mantle of 'boss', the so-called 'them', and those whom that boss has a responsibility for – in short, the 'us' of industrial relations.

During the 1970s, the decade I spent working under guvnors such as Biff Sands, Ben Tuke and Jim Cronine, I learned a lot about the differing characters and attitudes of 'real' Londoners – someone born and bred within the hard streets of the inner city – compared to those, like myself, born and bred in the industrial manufacturing north. In particular, I came to understand how the 'them' and 'us' relationship within the fire service differed between the two regions. My own theory rests on the better opportunities afforded within the capital and I will explain why.

Although back in the 1960s and 1970s there were reasonable employment prospects in a variety of occupations in West Yorkshire, the opportunities were nothing like as prolific as they were in London. It was this ready availability of alternative employment that I saw as having a strong bearing on the attitude displayed by individual firemen to those who had sought promotion.

When there are other means readily available to earn good money, it follows that men are not as insecure about their jobs; in fact, back in the 1970s, the duty system of whole-time firemen was such that following the two nine-hour day duties and the two fifteen-hour night duties, a man had three rota days off, allowing them opportunities to pursue alternative careers. Given the relatively poor wages of the fire service in those days, especially if a man had children and a wife to support, then the need for extra cash made working at two jobs to better the family's comforts a laudable pursuit.

Both in London and the provinces, no doubt because firemen have no qualms about working from ladders, 'being on the glass' (cleaning windows on a round) was traditionally a secondary source of income for those who needed it. Some men also earned a little extra by becoming pall-bearers, while others drove vans and trucks, or mini cabs and taxis.

In fact, there were more than a few London firemen who gave me the impression that they considered their job with the brigade as a 'nice steady earner' – it was a guaranteed wage, albeit a paltry one – but saw the taxi as the real source of their income.

Doing 'the Knowledge' was a gruelling pursuit but necessary for those who wished to obtain the coveted green badge that would allow them to ply for hire across the whole 620 square miles of Greater London. On average, it took more than three years, out on a moped in all weathers, to learn virtually every street and place of interest across the capital. Such determination breeds stubbornness, which in turn can instil strong opinions that, combined with earning more as a cabbie than as a fireman, can create tensions for the guvnor on a watch. In short, such men could become a bit of a thorn in the side of a guvnor who believed in running a tight ship.

The numbers of taxis on the many stations I served on as I was gaining my own 'knowledge' in terms of emergency work experience was proof positive of the determination held by all those who had managed to gain that coveted green badge.

Of course, there are exceptions to every rule: one of the most dedicated 'good hand' firemen I ever worked with was also a green badge taxi driver who had a huge brood of kids. The determination he had employed to gain that badge so as to better his family's lot at a time when the fireman's salary was so paltry was also deployed in his being a most competent member of the watch. Unlike some, his first love was being a fireman.

You might ask, what has all of this to do with supervising firemen? Well, it has to do with the nature of the 'them' and 'us' relationship.

In London, the relationship between the 'officers' (that is to say, the station/watch-based Leading Firemen, Sub and Station Officers) was far less stiff and tended to be more rough-and-ready, and I was sure that it was because of the 'second job' work mentality and the opportunities available in the capital.

Back in West Yorkshire, men would also work in between duties, but there were not the same opportunities as in London and firemen there seemed to me to more jealously guard their positions. As a consequence, the respect afforded even to the lowest rank of Leading Fireman was much greater than was generally the case in the Big Smoke. In Yorkshire, as the rank increased to Sub Officer and then to the white-helmeted Station Officer, the respect for that increasing authority was all too evident. True, as I have said, there were firemen who had lucrative moonlighting jobs, but there was always what I would call an automatic obedience, more akin to that found within the regular armed forces.

Across the board, mind you, the real distinction between 'them' and 'us' existed between a fireman and those of rank beyond 'guvnor' – that is to say, the Assistant Divisional Officer (ADO) and higher.

I have mentioned the great respect held by the grass-roots fireman on the shop floor towards the calibre of the extremely competent and experienced Tom Monsals of this world, but once the next rank came in – Divisional Officer (DO) – and certainly once a man had achieved principal officer status, such as Assistant Chief Officer (ACO), there was a real sense of division.

This wasn't manifested in disobeying orders, which could result in charges under the Fire Services Discipline Regulations, or in not being polite in their presence, rather in a bunker mentality, like the 'poor bloody infantryman' who sees his high-rank commissioned officer as occupying an ivory tower far removed from the real hazards of the front line.

I know of a former LFB colleague, a former Station Officer, well respected by his watch, who, after retirement, became a respected university academic. He wrote a PhD on the 'them' and 'us' thought process in the whole-time fire service. He concluded that many rank-and-file firemen considered that once a guvnor moved away from the fire station on promotion to ADO and higher, and was based in a headquarters office, he had committed 'treason' in their eyes.

I have experienced this mess-room thinking, where the collective of 12 or so front-line firemen create a view that no matter how competent or experienced a guvnor has been, once he moves up he cannot be seen in the same light, which is why the Tom Monsals of the brigade were quite a rarity in that they somehow better managed the psychology of their 'situations' and, as a result, were far more respected than was the norm, especially back in the periods to which my accounts relate.

Given this seeming demarcation, it is to me all the more impressive that whatever elements of unruliness or 'backchat' might have sometimes been present on the station, certainly the less busy ones, this was all forgotten when the chips were down and firemen and junior officers from fire stations were working in tandem with their senior ranked supervisors from headquarters to best ensure the safety of the public caught up in some fire or non-fire emergency.

For me, it was during the Moorgate tragedy of 1975, one of the brigade's most notable non-fire emergencies, that this cohesion, this *esprit de corps*, between all ranks was most evident.

Chapter 23

Moorgate, 1975

I was not in Carmel's good books when I set off for our first day duty on the fateful morning of 28 February. The previous night had seen me meet up with a few of the men from the station in west London to which I had first been posted following my transfer to the capital. Like on other nights in those first few months, we had ended up in the Hammersmith Palais ballroom or, to give it its proper title, the Palais de Danse.

In fact, positioned immediately next door to the busy local fire station where I had done several out duties, the Palais had often provided us some entertainment at throwing out time, with the brawls that broke out in the street below. Whether the scrap was the result of a jealous bloke, the girl he had been eyeing all evening having gone off with someone else, or some old grudge being settled, fuelled by the Dutch courage of drink, we didn't care; it was just amusing to watch, as we waited for the bells to clamour.

The excuse for my going out was the 21st birthday of one of the guys. With hindsight, I should not have gone. Carmel had a much closer relationship with her family and friends back in the north than I had with mine and, as a consequence, she had given up more than me in making the journey south. In fact, we had been squabbling quite a bit since the move and it was to do with the loneliness she was feeling in those early years when I was on night duty.

Although she had recently started a new job, she had not

developed any close friendships with other women and my long 15-hour nightshifts were a strain in more ways than one.

I also sensed it was getting to her that I was in much more danger when going on duty than I had been in my previous brigade. For me, facing this danger was easier, being part of a team: the camaraderie and banter, the black humour and the sheer force of the personalities of some of the crew made it easier to get through a shift. For those back at home, it is different. They are isolated.

Of course, she had appreciated before our move that to gain the experience I wanted it was necessary for me to serve on the busiest stations in the highest fire-risk districts. But appreciating something before the event is not the same as actually experiencing it.

But, like many young men planning a night out with like-minded friends, I had overlooked her overall feelings. I should not have been so callous.

As the night wore on, and that 'Oh, come on, another couple of pints are not going to do any harm' was put to me, I succumbed. My eventual arrival home at gone three after a taxi ride that had emptied my already meagre wallet saw me sleeping on the sofa.

Carmel was still fuming when I left for day duty the following morning. Such is the nature of temper, and the truth in the old saying, hell hath no fury like a woman scorned, that I set off without receiving the warm affection she was so ready to give in more harmonious times.

At 8.39 a.m., the usual crowd of commuters had clambered onto the train at Drayton Park, only a stone's throw from Highbury stadium, then home of the legendary Arsenal Football Club. Most of the passengers were employees of the banks, offices and other commercial premises that existed within the Moorgate and City district. The majority would have been regular users of the train, which only had three stops before Moorgate, where they would have made their way up to the thronging street to start their daily grind.

When the same train is taken on the same route every weekday, one becomes familiar with the train's speed and braking, and knows when the terminus is reached. In the case of Moorgate, it is a true terminus, with any further progress made impossible by the presence of a dead end tunnel some 50 feet beyond a set of buffers.

Station staff and some of the train's passengers would later recall that instead of the usual gradual slowing as it passed a certain point, the train began to accelerate. At 8.48 a.m., some nine minutes since its departure from Drayton Park and bounding along at perhaps thirty miles an hour, it smashed through the buffers, dug into a sand drag, then smashed into the dead-end tunnel. The massive impact caused years of accumulated encrusted grime to become dislodged, filling the air with a thick black veil.

The brigade's Control first received a call at 8.51 a.m. from the Metropolitan Police, who informed them that this was a major incident. Immediately on hearing this, fire appliances were sent from Barbican and Shoreditch, and these arrived within minutes. Barbican's Station Officer and crew rapidly descended to the affected platform, meeting a heavy surge of passengers exiting. Many had soot-stained faces, suggesting they had been within the greasy smoke of a smouldering fire.

On first entering the platform, the mid and central parts of the train looked normal. But the black veil in the air strengthened the idea that there was a fire. Only a year earlier, there had been an incident in a nearby station in which the heavy insulation on the cables supplying the traction current had ignited, and this seemed a similar scenario.

However, after speaking to a very shocked station staff, and having taken a closer look, the Station Officer realised with horror what had happened.

It appears that the first three carriages of the rush-hour passenger-crammed train had smashed into the dead-end tunnel. The massive impetus of what had been an accelerating train, combined with the great weight of the rest of the carriages behind, rammed the three carriages into the tunnel, which had space for only one and a half at the most. Such had been the massive pushing weight of the following cars that the second and third of these had reared upwards. The tunnel had once been used for larger cars, so there was more than the usual free space above the train and the tunnel roof. This had allowed the two cars to rear up. One of these cars was compressed to not much more than three feet.

The Station Officer in charge, with great coolness and presence of mind, at once sent a radio message initiating the Major Incident procedure, a transmission that few in the service are ever called upon to send, no matter how many years they might serve.

This puts into instant operation a well-drilled and practised contingency plan. Hospitals are alerted. Operating theatres are made ready. Medical teams are mobilised and a predetermined number of fire, police and ambulance appliances are dispatched.

The true scale of the disaster was unfolding as we arrived on the platform. Ambulance crews were on their knees, working urgently to resuscitate those passengers fortunate to have escaped the pulverising pressure of the first three cars, who were now prostrate and unconscious on the cold platform. Shirts and blouses were ripped, vests and bras removed to expose chests in their attempts to pump and force life back into them. As the ambulance staff set to work, the walking wounded staggered around in numbed shock at the surreal scene and mingled with those who had escaped any injury and who were rapidly exiting from the central and rear carriages.

The initial attendance crews had by now better evaluated the terrific task confronting them. Such had been the impact with the dead-end tunnel's structure that the front carriages had expanded sideways and upwards. The red-painted steel body panels were peeled back like the removable top on a can of corned beef. This had created a skin-slicing jagged edged steel barrier, blocking the already tight space between the train and the curved wall of the tube tunnel. It was clear that only the most slender of men would be able to squeeze past. Moorgate 1975 became one of the biggest challenges the brigade had ever been called upon to deal with.

Reinforcing appliances and crews soon approached from all compass points. It is at such emergencies that the good-natured rivalry and ribbing between stations and divisions, as mentioned earlier, plays such a huge part. That holding of 'bragging rights' as to which station responds to the most incidents, which units get on scene the fastest or which districts have the highest risks helps hone the operational efficiency of the whole fire force to its highest level. For when the chips are down, as here, deep below the capital's streets, with scores of people in peril, those healthy rivalries are subsumed; the fireman and top brass – 'them' and 'us' – attitudes are buried for the duration of the incident for the greater good of working as one huge and efficient lifesaving rescue team. No more was such a team effort needed than on that day.

Imagine, if you can, the scenario that confronted crews as they ventured with trip-hammer-beating hearts and adrenalin-fuelled nerves into the dark unknown. It was not the place for the faint-hearted. A claustrophobically cramped collection of compressed cavities of crushed, twisted and tortured metal. The stability of the wreckage was uncertain, as was that of the arched tube tunnel into which the train had impacted with such a mighty compressing force.

So, on top of the horrific thoughts of the carnage in a rush-hour train crammed with passengers, suddenly compressed to as little as a few feet, was the apprehension that at any second the wreckage might shift and collapse onto the rescuers.

Hand lamps only partially illuminated the awful blackness, but the lack of light and obscuring crush of wreckage could not silence the dreadful cries and moans of the trapped but still living, nor the certain knowledge that most within the front carriages would be dead.

At nearly every fatality attended, especially those in which the gender or age of the victim brought so sharply to mind my own loved ones, I had to fight very hard not to imagine them as the ones now dead.

Such emotions are heightened when the death scene is one in which an association with one's own kin exists, and so it was at Moorgate.

I had to remind myself that no one had forced me to leave the relative operational quietness of a provincial brigade for the hurly-burly busyness and potential hazards that are ever present within central London. I had made my occupational bed, but I would not be telling the whole truth if I said that on that first morning at this dreadful disaster there were not some moments when I wished I was not lying in it.

Such negative thoughts were compounded by the fact that Carmel and I had not been on our best terms when I set off for duty. I cursed my thoughtlessness at the callous behaviour I had displayed. Such is the nature of the human personality that when illness or accident strikes, arguments and disputes can be instantly forgotten in all but the most ruthless and vindictive souls.

And so, as the intensive rescue and retrieval operation got into full swing, the thought occurred to me, how many of those unfortunate victims, still alive but pinned by steel and timber, would forgive all wrongdoing to be freed from the carnage around them? And how many of the dying would, like soldiers grievously injured in battle, call out for their mothers or loved ones?

I am certain that I am not alone in such thoughts, though very few firemen would openly admit to them. The truth is that to do the gruesome side of the job, you have to have as genuine a wish to help others as does the dedicated nurse or doctor. Somewhere within that outer strength and toughness, a high sense of humanity has to exist.

It is that sense of humanity and public service that was the main driver in the rescue effort, as I looked around me at Moorgate, but I formed a conclusion that there was another motivator at work too, impelling those on the front line to work with an even greater sense of commitment than the very high level for which they were renowned. It is my belief that the close familiarity that most present had with the London underground provided a subliminal physical drive to the rescuers – a drive so strong that senior officers had to order some men to take a rest from their ultra-determined rescue endeavours. This ties in with what I have already said about the personal associations rescuers have with an incident. In this case, it was the iconic status of the underground, so familiar to all the emergency services, which seemed to push the workers on.

After all, who has not travelled by tube train? Who has not stood or sat as the carriages clattered, swayed, accelerated and slowed? Who has not scanned the coloured maps of the underground network or read the advertisements behind their glazing? Who has not studied the beauty of a smooth young face or the ravages of time on the elderly, and who has not been in the tight cram of rush hour forced close to the bodies of those next to you, some wafting the fragrance of a quality perfume or aftershave, others the sharp pungency of body odour?

Who has not gratefully secured the small folding seat that took one's bodyweight off feet tired and sore from a shopping or sightseeing trip? So glad to be sitting, even if your head is submerged in the swaying bodies of those around. Seats that were now lost, indistinguishable under the dead-weight wreckage of bogie wheels that had spun and rent asunder the carriage floor to sever limbs and suffocate those unfortunate to be in the wrong place at the wrong time.

So here the much-loved transport system had been, in an instant, transformed. Gone was the warm sense of workaday routine, replaced by the icy veil of instant death and carnage. The red-and-green tough fabric of seats was now pulverised by steel shards and panels, soaked with the blood of the masses. Those advertisements – for Liqufruta Cough Mixture, the Brook Street Bureau, to Visit Madame Tussauds and the London Planetarium – were now mangled and crazed almost beyond recognition under the twisted wreckage.

I spent the next day and the following two nights in that tunnel of death, but such was the tenacity and skill of all involved that all those who were still alive or could be extricated had been by the end of the first day.

All major incidents leave mental images; this was one of the biggest peacetime disasters handled, with 43 persons killed and 74 seriously injured, and hundreds left, if not physically damaged, with mental scars for life.

Those images were especially vivid for those who were close to the practical process of rescue and retrieval. My overriding memory, which still has a stunning clarity more than 35 years on from that tragic scene, is of that part of the carriage where the small folding seats were positioned – seats that Carmel and I had sat on one far away evening shortly after first coming to London and going into the West End, savouring our new environment. So thoroughly compressed were the leading carriages that it had taken two days of cutting the wreckage into slices and removing them to reach this particular portion of the train.

I had been detailed to make a check within this area and remember gingerly climbing into it. The risk of bacterial infection in the fetid atmosphere, heavy with that vile smell of human decay, was such that masks had to be worn. So pungent was the smell in that steel death vault that you thought it would never leave your nostrils; even up in the sweet air of the street, the pungent stench of putrefaction remained.

In a right-hand corner of the carriage, directly in front of those tilting seats mentioned, there was a bulging tangle of twisted steel; a convoluted crush of steel panelling, passenger grab rails and ripped, twisted seating. Protruding from the central and lower levels of this wreckage, in line with the position of the hands and lower limbs of an average-sized woman, my eyes focused on the waxy tone of human skin. First on the curving, shapely and muscled calf of a woman's leg, the foot still embraced by a high stiletto-heeled black patent leather shoe.

Two feet or so higher protruded the woman's hands. These were waxy, yellowy-green and delicate, and were held almost as if in a gesture of religious supplication.

Both legs and hands were unmarked of visible injury; it was only the time that had elapsed, along with the hopeless crushing under that shackling wreckage, that removed any hope that life could still have been present.

What has remained within the caverns of my mind, which on that first morning had been consumed with the guilt of my thoughtlessness towards my wife, was the glinting gold of a wedding ring – that band of gold, symbolising the endless unbroken circle of eternity, into which state that poor woman had been cast so violently on that fateful morning. Perhaps she had kissed her partner goodbye only a short time before boarding the train, the same man perhaps who had once carefully and lovingly placed that ring on her finger. Perhaps, like me, they had parted on less than amicable terms. I would never know the true situation, but these thoughts and that image remain undimmed.

Only a psychologist could explain whether the effect on the

personality of witnessing the tragedy of Moorgate 1975 played any part in the way my personal life subsequently turned out. Back then, working alongside men who had seen mates blown up before their eyes in theatres of war or in the London Blitz, one didn't whinge for fear of being seen as 'lacking in moral fibre' or of letting down the proud name and reputation of the brigade. No, at that time there were few, if any, counselling or defusing services. A man had to steel himself and develop a mental callus to shield him from the raw realities of the work. Otherwise he should capitulate and apply for a job in a less emotionally demanding arena.

Chapter 24

Death on the Mile End Road

Two quick rings of the call bells by the Sub Officer signal the end of the hour-long dinner break, in which Mary, the daytime cook, has weighed us all down with stew and dumplings, mashed potatoes, carrots and peas, followed by a portion of bread-and-butter pudding almost as big as a house brick.

We are on the second of our two day duties. The previous day's 9-to-6 shift had been uneventful and, save for the usual AFA false alarm and one other infuriating mickey, the only action seen had been a rubbish fire in a derelict house that had been dealt with in 20 minutes. Yesterday had been our day for carrying out standard tests.

Virtually every single item and piece of operational equipment is subjected to a testing regime at various periods, dependent upon what it is used for. Some tests are carried out daily, some weekly, whilst others are done monthly, quarterly or annually.

It is reassuring to know that when the chips are down it is highly unlikely that equipment will fail.

We are scheduled for drills after lunch, and few are relishing running up and down ladders laden as we are with Mary's fulsome meal. It is with a collective sigh of relief that the grey skies on this winter's day finally begin to release the rain that has been threatening all morning.

Safety requirements dictate that we normally don't drill outside if the weather is inclement and so we find ourselves in the mess room to participate in 'lectures', which can cover virtually any subject related to our work.

'Right, listen in, you lot,' barks Sub Officer Roy. 'We can't practise our firemanship outside, so we'll practise some here. Practical firemanship will be our session for the next hour or so. Let's take a look at ventilation at fires.'

'Why do we ventilate . . .' – the Sub looks round him, deciding who to pick to answer his question – '. . . Jim?'

'We all want to ventilate Jim,' jokes Niall Peating. 'To release some of the guff in his head that he's always spouting.' Niall is a one-time schoolteacher with an Honours degree in economics.

'Well, Niall, releasing is one part of the answer, but I'm not so sure yours is the correct full answer,' retorts the Sub. 'Ignore 'em, Jim, and carry on, if you will,' he continues.

'Right, Sub. Well—'

As is often the case, the call bells cut someone off mid-sentence.

'You jammy sod,' quips Niall again. 'I bet you didn't know the rest of the answer,' he laughs, as Jim gives him a Victory sign while running to the pole drop.

John Lydus is duty man for the shift and he is reading the teleprinter message as we plunge down into the appliance room; the green indicator light above the huge red station doors is already lit. '*RTA* [Road Traffic Accident] – *persons trapped – Mile End Road. Pump only*,' he shouts.

Within 20 seconds, the Pump is thundering and blaring along roads made greasy by a cold drizzle, weaving in and out of the heavy traffic – a regular feature of this major artery into the capital from the east, once the busiest dockland route in Europe.

A fair proportion of the vehicles are lorries on the road transporting a wide range of goods into and out of London, and it is one of these, a refrigerated articulated truck, en route to Smithfield Meat Market in Clerkenwell, that has collided with a private car at one of the many junctions controlled by traffic lights along this major highway.

The two-tone horns echo from the high walls of the densely packed buildings alongside and the metallic, tractor-like rattle of the diesel engine surges up and down as we hang on in the rear, swaying like sailors in a force nine gale.

The radio crackles into life: '*From Sub Officer Devine at Mile End Road – one saloon car in collision with articulated lorry. Three persons trapped. Efforts being made to release, ambulance required.*'

'That don't sound so friggin' healthy, fellas. Get your high-visibility surcoats on,' shouts Ben Tuke above the engine's roar and the cacophony of the air horns.

As the Pump halts and parks in a position ahead of the lorry and car, so as to 'fend off' approaching traffic, we can see that the car has gone at least a half of its length under the semi-trailer of the lorry.

On the basis of having attended similar incidents in the past, I reason that what we might see within that crumpled wreckage of the car will not be pretty at all, especially after our heavy midday meal.

I jump down from the Pump and notice that a young policewoman is doing her best to direct the traffic around the collided vehicles, and that the traffic lights are not working.

Not all articulated lorries at that time had side crash rails and it looks as if the absence of these on this vehicle has meant that the car has gone straight under the trailer. The severe impact has caused the roof of the white saloon to split and tear back.

'What's the score, Dolan?' enquires Ben Tuke of Sub Officer Devine, who is on a temporary secondment and in charge of the local station ground's appliance.

'It's a nasty one, guv. There are three trapped, from what we can make out at this stage,' he replies, deliberately keeping his voice low in case any of the casualties are able to hear.

'We've laid out a hose reel in case anything lights up and disconnected the lorry's batteries. I've sent an informative, plus a request for an ambulance, but for sure we are going to need the help of the ET [Emergency Tender] on this one. I've got two men under the trailer, trying to get into the front of the car,' he continues, his speed of delivery indicating the urgency within his demeanour.

'Well, the ET should be here in a jiff with the heavy cutting gear and jacks. Let's see if we can get into the rear seats whilst we can, Dolan, eh?' says Ben Tuke, before shouting over to Niall Pointer and Geoff Joynt, 'Get the crowbar and try to force the rear doors, pronto, and see what you can do for anyone in the rear, yeah. Andy and Jim, you look after the front of the car.'

The lorry driver – a short, wiry man with a tight crew cut, wearing a black T-shirt and trousers, and with a deathly white face – is leaning on a round-topped steel bollard on the pavement. It is one of those old-fashioned constructions placed to protect pedestrians should a vehicle mount the pavement. Its round top has been burnished from a lifetime of wind, rain and the resting hands of many passing folk. It is now serving to steady his hands, which, as a result of shock, are trembling like an aspen leaf. A grizzled, snow-haired policeman is talking to him, while writing notes in his black notepad.

The roof above the rear passenger compartment is twisted back hideously. The grey-coloured cloth lining is splattered with blood, which, in its absorption by the cloth, had changed colour from claret red to a bright brown. A woman, who appears middle-aged, is on the rear seat, but her left leg is twisted grotesquely around the rear of the front passenger seat, and her right leg bent sharply under her downward facing form.

Again, it is the sight of the everyday, commonplace items present in the vehicle that adds such poignancy to the scene: the partly used cardboard box of tissues in the door pocket; the upturned handbag on the floor, burst open with the impact, scattering lipstick tubes, mascara brushes and perfume bottle; a toy mascot hanging in the smashed windscreen; the lucky dice that have brought no good fortune. A child's doll strewn about, almost a miniature of the full-size human forms caught up in such unfortunate straits.

The white paint on the inside metal of the car doors is streaked with the bright red of blood, now slowly congealing and meeting up with a darker red pool running from the front half of the car, which is embedded under the trailer. As I thought, this is going to be a nasty job all right.

Our guvnor bends his tall, bulky frame over the rear of the car. 'How's it look, Niall?' he asks quietly.

'It ain't good, guv, but there's a faint pulse in the neck of the woman in the rear. We've got to get her out quick.'

The strident blare of the siren on the Emergency Tender heralds its arrival, and on its tail is the ambulance.

Under the lorry's semi-trailer, the squat form of Jim Peating and the much taller and slimmer figure of Andy Carlisle are struggling to reach the two casualties in the concertinaed front part of the car. Andy's hand lamp beam scans the distorted, twisted metal and the heavily blood-stained upholstery. A fireman from a northern brigade, he had served on stations that had attended many RTAs on one of the country's busiest motorways. He instinctively knows that the damage to the car and the severity and position of the severe impact is such that it will be a miracle if anyone gets out of this carnage alive.

I am crouching under the trailer, next to Jim Peating, my boots in a small pool of what looks like engine coolant or brake fluid that had been released on impact.

'How's things, Jim?' I ask.

'It's bad, mate, real bad. Andy says we'll need to get the trailer jacked up by the ET crew to get into the front of the car. The poor fuckers in the front must be in a right mess,' he replies, shaking his head, from which he has, like the other two of us, removed his helmet to enable him to work in the contained space. I crawl out from under the trailer to update the guvnor on the situation and as I stand up, a sharp odour hits my nostrils. It is a smell I have experienced before at other road traffic incidents and on those gory occasions when dealing with suicides on the underground rail network. A vile, nauseating whiff, it comes from not only the human blood so often spilled in these sorts of emergencies or the smell of twisted metal, a bit like that of a burning clutch plate, but also from the stomach-heaving, throat-retching stench of the involuntary opening of a casualty's bowels and bladder.

As I am informing the Station Officer of the need for the trailer to be jacked up, the large white saloon of the duty ADO pulls up to the kerb about 20 feet behind the incident, its blue beacon spinning and then stopping as the car parks up.

Tom Monsal has been informed of the incident by Fire Control. Brigade policy dictates his supervisory attendance at 'persons trapped' incidents of this type. He heads towards our guvnor, who is detailing the Leading Fireman in charge of the ET as to what will be required of them.

The eyes of 30-year veteran Tom Monsal scan the scene. It's immediately apparent to him that this is likely to be quite a protracted job and therefore he will be inwardly reassured that the experienced Ben Tuke, a man with whom he has worked alongside for years as a Station Officer himself, is on the spot.

Having been appraised of the situation by Tuke, Monsal responds, 'OK, Ben, I've got the picture. I can see you have the job well in hand. I'm not going to formally take over, but I'll be around. I'll get rigged in fire gear now.'

Within a minute, Tom Monsal is rigged in his operational uniform, normally neatly stowed in the large boot of his car. He removes his helmet before crawling under the semi-trailer where Ben Tuke is on his knees, his huge form such that even in this position his back is almost touching the underside of the trailer's bed.

Each appliance carries a box lamp and two of these have been positioned under the trailer, their yellowy-white beams illuminating the tangled wreckage of the car and reflecting off the white paint of the bodywork, which has been smashed like an eggshell.

The ET crew, specially trained in extrication techniques, are now under the trailer, assessing the situation and, through experience borne from hundreds of incidents, have soon formulated their plan.

'We'll need to jack and block the trailer, guv,' the Leading Fireman remarks, his advice directed to both the ADO and Station Officer. 'First, though, the articulated tractor unit will need to be uncoupled from the fifth wheel and service cables,' he goes on.

'Of course, but let's get it done like yesterday,' instructs Tom Monsal. 'We've three trapped and, if there's to be any chance for any of them, we've gotta be sharp. So go to it.'

Before the jacking up begins, one of the ET crew deals with the uncoupling of the tractor from the trailer. The police have already done their photographs, marked the surface of the Mile End Road for evidential purposes and taken the 'spy in the cab' tachograph from the lorry's cab, so the tractor can be moved enough to permit the trailer to be raised. Within no more than 15 minutes, the trailer is being lifted, inch by inch, the groans and squeals of impacted metal sounding as the weight of the trailer is taken slowly off the car. Whilst this is taking place, the ambulance crew plus two doctors and a sister from the casualty department of the nearby Royal London Hospital in Whitechapel are attending to the grotesquely contorted rear-seat passenger.

Displaying his customary presence of mind, Ben Tuke had earlier sent a priority radio message requesting a medical team. If the passenger can be saved, they are the ones to do it. As the weight finally comes off the car, the rear passenger, who has now been extricated, is rapidly conveyed to the nearby hospital.

Now that the trailer has been raised clear of the car, the car can be slowly pulled back the eight or so feet it has travelled under the trailer. Over an hour has gone by since we first arrived, and it is getting dark now; the tall lamp standards alongside this major East End highway are flashing into their sodium-yellow light, casting long shadows around the severely smashed structure of the car. Now that the weight of the trailer is removed, we can see that the front roof of the car is almost flattened to steering wheel level.

It takes another 45 minutes or so of careful cutting before the tortured, twisted, tangled metalwork can be removed. Throughout this time the second ambulance waits to the rear of the scene, its bonnet already pointing in the hospital's direction, its rear doors wide open, the white interior lights revealing the familiar red blankets of the London Ambulance Service.

Unfortunately, it soon becomes apparent that there is no need for an urgent blue beacon and braying siren for the dash to Whitechapel's famous infirmary. As we cut through the layers of steel, walnut fascia, rubber, glass and velour cloth, we uncover a ghastly scene of carnage in which death would have been instantaneous. Instead, the ambulance becomes a respectfully slow conveyor of death.

Only an hour or so earlier, the driver and front-seat passenger had been talking, feeling, thinking human beings. Now they are as lifeless as the cargo of beef inside the trailer under which the unfortunate car had impacted with such a shocking severity.

As we head back to the station, we know that there was little any of us in attendance at this dreadful job could have done to improve the situation, save for releasing the rear-seat passenger, who, we are informed later, did survive.

The Fire Service exists to save life whenever it is humanly possible to do so and these non-fire 'special services' have become an increasing part of operational work. Our natural sense of loss and dismay at the waste of human life is apparent when we hear later that the accident was caused by the car shooting the traffic lights, but our sense of the futility of needless fatalities is never lessened, no matter how many fatal incidents we attend.

Chapter 25

Fire in the East

London's East End had been a magnet for immigrants for many years when I first set foot within its mean streets. Successive waves of immigrants had been arriving for many decades, populating the areas around Whitechapel, Shadwell and Spitalfields. Flemish silk weavers, Irish peasants chancing their arms rather than succumbing to starvation in a potato-blighted land, Jews escaping Russian pogroms; then, in the late 1960s, many arrived from the Indian subcontinent in the wake of political upheavals and tensions there. Each cultural group brought its own identity with it, the variety of religions and their associated rituals and festivals, all of which strengthen and commemorate cultural belief systems and values, adding a wide range of colours to the overall greyness of so many of the East End's streets.

The local immigrant population also provided for those of us crewing the local fire appliances a rich vein of human interest and, all too often, human drama. Within those often mean streets, stretching from the Isle of Dogs up through Poplar, Shadwell, Bethnal Green, Whitechapel, Shoreditch and Hoxton, we were frequently forced by a vile smoke and searing heat to crawl, nose to the floor, along lengthy corridors as we moved towards the orange glow of a fire.

The traffic coming out of central London on the Farringdon Road, abutting the Square Mile, is even heavier than usual, as I drive my trusty old car the 15 miles to the station. At least I am heading against the stream of vehicles, I think, as I weave amid the ever-present black taxi cabs (some of which might be driven by off-duty colleagues) and the dispatch riders – daredevils with a 1,000CC-powered nonchalance between their leathered limbs.

The stop-start journey allows me to take in the blackened, fire-ravaged hulk of a former office to my right. Along with over 50 crewmates, I had tackled that inferno only a month back. More and more jobs at the time were being attributed to arson. London was seeing the types of fires New York had been seeing for years: incidents where fires were ignited out of malice or revenge, for insurance fraud or to hasten the demolition of buildings to create land to satisfy the insatiable demands of property developers, a constant feature in the major cities.

As I make a series of back-doubles en route to the station that cold early winter's evening, I look up at the ink-black, star-spangled sky that heralds a keen frost and wonder what lies in store for us on the night duty. Our watch had been unusually busy over the previous month.

Things are different tonight, though. By eight o'clock, when we settle down to our evening meal, we have so far only received a call for the Pump to attend what turns out to be yet another frustrating malicious false alarm. This means we can enjoy a delicious curry prepared by 'chef' Geoff Joynt. The sweet aroma of spices and exotic cooking have become features of life within the inner East End, now that an increasing number of Asian immigrants have chosen to take up residence in the area. A sure case of East meets East.

Like most who have served a decent time on the busiest inner-city stations, I have developed an ability to feel what is almost a premonition about the next serious fire. But perhaps it is no more than my subconscious working out that, by the law of averages, a city centre sharp-end station will be bound to pick up a 'real goer', as we call it, sooner or later. Or that, given the frosty night, the likelihood of a serious blaze involving residents will increase since portable heaters will be brought out to counter the bone-chilling temperatures. Calls to residential premises in the early hours on such a night will always create a dark sense of foreboding within the crew.

The night drifts by without further occasion, so I am relaxing in the early hours, allowing my mind to wander. Many a time during the small hours I found myself asking the question, what is it within this down-at-heel quarter that I find so absorbing? I certainly held an affection for it that was hard to define: it was a little like that held by a child for its favourite rag doll – at once comforting but, at the same time, slightly offbeat.

I only had to close my eyes to bring back the countless memories of responses made to fires within the locality. Always, especially in the night, surrounded by the tightly packed commercial and residential premises, there was the great sense of anticipation and uncertainty when the call bells rang – would this be the fire to end all fires? The one from which you might not return?

I could recall the heavy banked-down smoke of hundreds of calls, its density obliterating street lamps. I could see the 'Proto' BA crews, goggles on foreheads, breathing bags engorged, awaiting the instruction to enter a deep basement full of life-taking smoke; or the unconscious form draped like a stole across a back of a fireman, him deftly stepping down the escape from 40 feet up. I could hear the sounds of a burning building and the crash as the fire goes through the roof, releasing the bottled-up heat and smoke and, in its freedom, rising at high speed to form a swirling, roaring, crackling column of flames and embers, looking like a million fireflies shooting towards the night sky.

I could recall the fires in tenements, sweatshop factories, dockside warehouses . . . those 'never to be forgotten' blazes where, in spite of all of our best efforts, lives had been lost. Lives of men and women of all ages, but most remembered were the tragic deaths of youngsters, plucked from life and their parents without the chance to say goodbye.

Often as we returned from a call in the early hours, I would look around at the mean and moody streets and never fail to be intrigued by the creatures of the night, people out and about at the most ungodly hours. Not all of the women would be on the game, though some were. And while some of the men on the streets might well be seeking their services, what about the others? Some might be on their way home from nightclubs and gambling dens, or making an early start as an office cleaner or market porter, like Ricky Tewin once had. Perhaps some of those populating the seedy streets were suffering from mental illness, possessed of a troubled mind to which the healing balm of sleep will not come.

Was it the mysterious nature of the East End that captivated me, with its melting pot of cultures and history? Flicking through my memory bank, I get lost in my reverie, until out of nowhere comes the sudden illumination of the station lights, followed a fraction of a second later by the urgent toll of the bells. It is just before 4 a.m. The call slip indicates multiple calls to an incident in a district infamous for fatal fires.

Our Pump Escape and Pump are on scene within four minutes, just in time to witness yet again the horrific spectacle of a person plummeting from four floors up, their clothing ablaze, to land with a sound like a huge melon splitting as the body hits the pavement.

'We've got another bastard of a job here, boys,' exclaims Bert Braxton breathlessly, as we rapidly dismount.

Ben Tuke instructs Ricky Tewin and me to check out the rear of what turns out to be a five-storey period house. The sight that greets us almost transfixes us. High up on the steeply pitched roof we can see at least three people clinging to the ridge plate. They are clad in the long shirts, known as kurta, fast becoming a familiar site in this Asian ghetto. It would seem they have escaped from the top floor up onto the roof to escape the awful flames and smoke issuing from most windows on the third and fourth floors.

'We'll need to get up there to calm them, else there could be others ending up like the poor sod at the front,' I blurt out to Ricky.

'Yes, mate, but how are we going to reach 'em with all that fire?' he breathlessly replies.

'Hook ladder and the Dewhurst pitched there on that second-floor window on the gable end, and then to that parapet wall above the roof pitch,' I reply, pointing towards the blazing building. I know that the escape will reach the roof from here, but it's already in use, providing a covering jet at the front, and there might be too long a delay before another escape gets here.

The wails and cries for help from on high mingle with the drone of the pump feeding the jets already at work. In the distance, the urgent braying of two tones heralds the rapidly nearing reinforcements that Ben will have ordered. As I dash back to the front to apprise the guvnor of our rescue intentions, the radio metallically sounds through the cab speaker: '*Dwelling house of five floors, 25 feet by 50 feet. 50 per cent of ground and first floors, 75 per cent of second and third, and whole of fourth floor alight. Unknown number of persons involved. Building being searched by BA crews. Escape and covering jet in use. One person jumped from upper level before arrival, injured, awaiting removal.*'

I inform Ben of the situation.

'Right, do that and grab somebody to help pitch the Dewhurst,' he replies. 'Get 'em as secure as you can until we knock this bleeder down. And for God's sake, mind how you go.'

The Dewhurst is soon pitched and in seconds Ricky, with a rescue line in a rucksack-like bag on his back, is ascending it to a position where he can smash open the sash window at the gable end. Using his personal axe, he soon has the glass out of the window, at which point a lot of smoke vents and then lessens. I bring up the hook ladder and push it up to him.

He grasps it after taking a leg lock through the rungs and is soon astride the sill. From there, he punches the ladder up to the projecting parapet wall which we had identified earlier. He twists the ladder and hangs the two-foot serrated bill onto the brickwork. It's now or never and, almost as if to confirm this urgency, we hear the further cries of those trapped on the roof through which the fire might burst at any second and envelop them.

We are in luck: the ladder bites, in spite of it being on brick instead of the wooden sill of the fire station drill tower, and Ricky precariously transfers his light weight onto it. In a few seconds, he is on the ridge.

Later, Ricky told of how he had slowly inched onto the ridge plate and how horrified he was on seeing a lot of nasty yellow smoke pushing up through the defective mortar of it, indicating great heat and pressure from below.

Back at ground level, ADO Tom Monsal has just arrived.

'What we got, Ben?' he enquires, having located our burly boss halfway up the first-floor landing.

'It's a nasty one, Tom. I think the place houses a fair few Asians, all from one family. There are several trapped on the roof and I've got three fellas using a Dewhurst and hook ladder, trying to get to them.

'As you can see, we've an escape and two jets at work, and there are two BA crews searching. And we have a jumper, which I guess you can't have missed on your way up here.'

'You're right, Ben. It's a bad job,' Monsal exclaims, signalling to the staff officer who has just arrived from the Divisional HQ.

'Sub, send me a "*Make Pumps Six, one to be a Pump Escape, plus one Turntable Ladder required*",' the ADO instructs.

'We'll be on the safe side with those extras, in case we have problems getting those on the roof down,' Monsal advises Ben Tuke.

'Got it, boss,' Ben replies, stepping back into the acrid smoke where two men are crouched low, working a jet slowly into the inferno.

Ricky is now on the roof and no doubt his sudden appearance over the end wall has caused a simultaneous sense of fear and salvation for the three Asians clinging for their lives to the side of a crumbling brick chimney stack.

Fortunately, one of them speaks English and soon Ricky has passed on the message that they must all keep calm and that the brigade will get them to safety, but they should continue their tight hold on the chimney stack.

The colour of the smoke and the speed of its percolation indicated a fire about to burst through, but the last thing Ricky wanted was for them to sense this. If they did, the resulting panic could see three others lying on the hard pavement far below, preferring the option of a broken back to being burned alive.

I have taken the large searchlight from the Pump and its powerful white shaft is illuminating the smoke-wreathed roof far above.

Taking the line container from his shoulders, Ricky carefully secures the line to the stack. Then he feeds the line out, getting his interpreter to instruct the others to take a hold with one hand whilst still grasping the ridge plate or chimney stack with the other.

It seems an eternity since Ricky made it onto the roof, whereas in truth it is no more than ten minutes.

'Hang in there, Ricky,' I bellow. 'There's probably another Pump Escape almost here.'

In the meantime, the ADO and Ben have ensured that three BA teams are now totally committed to the search and firefighting operation. The intense heat indicates yet another accelerant-fuelled arson fire. I think back to my premonition about our picking up a real nasty job.

The extra Pump Escape is now on scene and within minutes the three on their decidedly dodgy roof perch are safe on terra firma, where they all bend and kiss the cold pavement.

Seeing us, Ben comes over and congratulates us on our efforts, before detailing us with another task.

'You guys need to assist in the search effort. We have suppressed the fire enough to work without BA. It's still very hot up top, but bearable, so go to it,' he instructs, adding, 'Don't disturb things because it looks like an arson job.'

The BA men have already located two men and two women unconscious in a rear room on the first floor and carried them down to the waiting ambulance crews. It has been extremely hot and extremely smoke-logged inside all floors, so it looks touch-and-go as to whether those four will make it.

The heat is still high as we enter and gets hotter the higher we go; soon, the sweat is running fast under my tunic and leggings as we ascend the charred, blackened and weakened staircases, keeping close to the walls where, hopefully, the treads are the strongest. It is not long before we sense a stomach-churning odour – that peculiar sweet smell of burned human flesh. The BA teams would not have detected this with their sets on.

I scan the piles of shattered, smoking lath, plaster and timber and my light picks out the grotesque sight of two badly burned bodies: blackened hulks, looking like a huge piece of charred timber spar, but in reality the remnants of what only an hour earlier had been living people.

The ADO sends an informative, recording the details and locations of the two deceased, then meets us on the stairs to witness the victims with our guvnor.

'Leave everything as it is, lads. The brigade photographer is en route. I will need statements for the Coroner from you two who discovered these poor buggers,' Monsal instructs.

'Push on now, up to the top floors, as quick as you can. I sincerely hope I am wrong, but if there is anyone still inside I don't hold out much hope for them, I'm afraid, given the temperature and smoke that will have been at this highest level.'

It is only after we force open a heavy panelled door adjacent to the head of the stairs which the fire must have raced up as if a chimney flue that we wish Tom Monsal had been wrong as to there being no hope. Inside a room of no more than about ten by twelve feet, with walls not burned but heavily blackened by the dreadful dragon's breath of smoke, there is a double bed positioned alongside a sash window. Lying on her back is the lifeless form of an Asian woman of about 30, the beautiful features and high-cheekbones so prevalent amongst her race still preserved. Her raven-black long hair is draped across the once-white but now smoke-stained pillowslip, and from her left nostril I can make out a gossamer thin small sphere of mucus, probably the last breath she had taken.

As I take in the scene through the still lingering smoke and oppressive heat, my gaze lowers down the sooty crumpled sheets. My eyes lock onto two figures. It is an image forever etched into my mind. Kneeling on the floor alongside the bed are two small girls of no more than five or six. Their frail arms are encircling their mother's thighs, whose own left arm is holding them both, as if trying to shield them from the horror that has befallen them. Other than a slight smoke blackening of their natural skin colour, their lifeless forms appear undamaged. At this precise moment, I can only consider that the heavy door has protected them from the flames and they have died by asphyxiation.

In all of my many years of experience, I have been unable to completely case-harden myself when faced with such images. It is almost impossible not to be affected by the loss of human life and such emotions intensify whenever children perish. No matter how well our minds try to shut out the unpleasant, there remains within the deepest subconscious a file of images that can resurface in future dreams or reflections.

Thankfully, in my years with the Fire Service such tragic scenes were tempered by the many occasions when life was saved. But perhaps it is the starkness of the worst sights that remind us of the sword of Damocles.

These thoughts seemed to be a bedrock from which flowed a powerful current of tenacity and resolve. For me, and I know for others too, it was this that impelled us to continue to hold our vigil over London, responding unquestioningly to whatever emergencies arose.

On reflection, I feel privileged to have been a part of that vigil, within inner London in general and the East End in particular. You see, so much of human life is there. It is present in abundance within that diverse and cosmopolitan environment, just as it was a constant on the fire station amidst the raw humanity of trusted crewmates, men who were always capable of ready banter, which in turn generated a close camaraderie.

Back then, we were able to gain an immense job satisfaction from serving on the front line within the highest-risk districts during such busy decades. Those were thrilling years, in which our developed instincts for saving lives and property could be unleashed without inhibition. How different from what arose in subsequent years, when the unintelligent application of what are essentially noble concepts of health and safety became disproportionate to the extent that the safety of rescuers became of seemingly greater importance than that of those awaiting rescue.

That can never be right. Such an unbalanced approach corrupts the noble business of the fire and rescue service and taints the fine memories of all who made the ultimate sacrifice whilst striving valiantly to protect those in peril.

Acknowledgements

I would like to say a big thank you to the following for the part they played in allowing my story of fire and rescue work in London during the 1970s to be told.

At Mainstream, I am indebted to Bill and Peter, for acquiring my manuscript in the first place; to Debs Warner, for her skilful and incisive editing; to Graeme Blaikie, for pulling so many things together; and to Fiona Atherton and Francesca Dymond, for their publicity involvements.

I also want to say a big thank you to my agent, Robert Smith, for his initial recognising of the potential within my accounts and for his professional advice and support, which have played such a part in bringing this book to fruition.

Last but by no means least, a big thanks to Anne and Mark, who have supported my endeavours over a lengthy period and shown much patience during the long hours I spent away from them, putting these accounts together.

I have quoted lines from a poem I read in the brigade magazine during the 1970s and want to give thanks to the author, now deceased. The poignancy of the lines remained with me for many years and I have tried to recollect them as best I can, having read the poem itself nearly 40 years ago.